T0265350

Lectures on
Lie Groups
Second Edition

SERIES ON UNIVERSITY MATHEMATICS

ISSN: 1793-1193

SERIES ON UNIVERSITY MATHEMATICS – VOL. 9

Lectures on
Lie Groups

Second Edition

Wu-Yi Hsiang

University of California, Berkeley, USA
Hong Kong University of Science and Technology, Hong Kong

 **World Scientific**

EW JERSEY · LONDON · SINGAPORE · BEIJING · SHANGHAI · HONG KONG · TAIPEI · CHENNAI · TOKYO

Published by

World Scientific Publishing Co. Pte. Ltd.

5 Toh Tuck Link, Singapore 596224

USA office: 27 Warren Street, Suite 401-402, Hackensack, NJ 07601

UK office: 57 Shelton Street, Covent Garden, London WC2H 9HE

Library of Congress Cataloging-in-Publication Data

Names: Hsiang, Wu Yi, 1937–

Title: Lectures on Lie groups / by Wu-Yi Hsiang (UC Berkeley & Hong Kong
   University of Science and Technology, Hong Kong).

Description: Second edition. | New Jersey : World Scientific, 2017. |
   Series: Series on university mathematics ; volume 9

Identifiers: LCCN 2017003517| ISBN 9789814740708 (hardcover : alk. paper) |
   ISBN 9789814740715 (pbk. : alk. paper)

Subjects: LCSH: Lie groups. | Lie algebras.

Classification: LCC QA387 .H77 2017 | DDC 512/.482--dc23

LC record available at https://lccn.loc.gov/2017003517

British Library Cataloguing-in-Publication Data

A catalogue record for this book is available from the British Library.

Typeset by Stallion Press

Email: enquiries@stallionpress.com

Printed in Singapore

Preface

In this rather short lecture notes, we try to provide concise introductions to selected topics on the important basic and also the most useful part of Lie group and Lie algebra theory, highlighting the major achievements of Lie, Killing, Cartan and Weyl.

As it is traditionally well-known, probably due to the remarkably early appearance of the wonderful classification theory of semi-simple Lie algebras of Killing–Cartan, which consists of an extremely impressive "tour de force" of linear algebra techniques, most books on the theory of Lie groups and Lie algebras as well as on symmetric spaces are essentially following the original steps of Killing–Cartan and hence often become mainly algebraic, consisting of long sequences of intricate, refined techniques of linear algebra. Somehow, even the beautiful geometric-algebraic duet of Cartan theory on Lie groups and symmetric spaces was somewhat obscured by the burden of algebraic technicalities. I think such presentations often make such a wonderful and useful theory quite difficult to learn and then to apply, say, for graduate students of mathematics or physics.

In retrospect, with the benefit of hindsight, the theory of compact connected Lie groups turns out to be the *central part* of the entire theory on semi-simple Lie groups and Lie algebras, as well as the theory on Lie groups and symmetric spaces on the one hand, and on the other, it is also technically much simpler and conceptually much more natural to develop a properly balanced combination of geometric-algebraic approach. Anyhow, this is the route that we are presenting in this short lecture notes.

We note here that the orbital geometry of the adjoint actions of simply connected compact Lie groups that we discussed in Lecture 3, in fact, already constitutes the central core of the entire structural as well as classification theory on semi-simple Lie groups and Lie algebras, as well as that of symmetric spaces. In the compact theory, the Frobenious–Schur character theory of linear representations, the maximal tori theorem of É. Cartan and Weyl's reduction fit beautifully, while the determination of the volume function of the principal orbits (i.e. of G/T-type) of the adjoint action naturally leads to the remarkable Weyl's character formula (cf. Lecture 3), which achieves a far-reaching improvement of É. Cartan's highest weight theory on representation of semi-simple Lie algebras.

In this lecture notes, we provide a concise, naturally developing discussions of the compact theory in Lectures 1–6, and then treat the Killing–Cartan semi-simple theory as a kind of natural extension of the compact theory in Lectures 7–8, the beautiful duet of Lie groups and symmetric space (i.e. Cartan's monumental contribution) as the concluding Lecture 9 of this series of selected topics on Lie theory. It is the sincere hope of the author that such a presentation of the basic core part of Lie group theory will make it easier to understand and to apply Lie theory to study geometry and physics, etc.

Contents

Lecture 1

Linear Groups and
Linear Representations

1. Basic Concepts and Definitions

Definitions 1. A topological (resp. Lie) group consists of a group structure and a topological (resp. differentiable) structure such that the multiplication map and the inversion map are continuous (resp. differentiable).

2. A topological (resp. Lie) transformation group consists of a topological (resp. Lie) group G, a topological (resp. differentiable) space X and a continuous (resp. differentiable) action map Φ: $G \times X \to X$ satisfying $\Phi(1, x) = x$, $\Phi(g_1, \Phi(g_2, x)) = \Phi(g_1 g_2, x)$.

3. If the above space X is a real (resp. complex) vector space and, if for all $g \in G$, the maps $\Phi(g) : X \to X : x \mapsto \Phi(g, x)$ are linear maps, then G is called a real (resp. complex) linear transformation group.

Notation and Terminology 1. A space X with a topological (resp. differentiable, linear) transformation of a given group G shall be called a topological (resp. differentiable, linear) G-space. In case there is

1

no danger of ambiguity, we shall always use the simplified notation, $g \cdot x$, to denote $\Phi(g, x)$. In such a multiplicative notation, the defining conditions of the action map Φ become the familiar forms of $1 \cdot x = x$ and $g_1 \cdot (g_2 \cdot x) = (g_1 g_2) \cdot x$.

2. A map $f : X \to Y$ between two G-spaces is called a G-map if for all $g \in G$ and all $x \in X$, $f(g \cdot x) = g \cdot f(x)$.

3. A linear transformation group $\Phi : G \times V \to V$, or equivalently, a homomorphism $\phi : G \to \mathrm{GL}(V)$, is also called a linear representation of G on V. Two linear representations of G on V_1 and V_2 are said to be equivalent if V_1 and V_2 are G-isomorphic, namely, there exists a linear isomorphism $A \colon V_1 \to V_2$ such that for all $g \in G$ and all $x \in V_1$, $A \cdot \Phi_1(g, x) = \Phi_2(g, Ax)$, or equivalently, one has the following commutative diagrams:

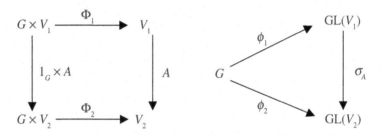

where $\sigma_A(B) = ABA^{-1}$ for $B \in \mathrm{GL}(V_1)$.

4. For a given G-space X, we shall use G_x to denote the isotropy subgroup of a point x and use $G(x)$ to denote the orbit of x, namely

$$G_x = \{g \in G : g \cdot x = x\},$$

$$G(x) = \{g \cdot x : g \in G\}.$$

It is clear that $G_{g \cdot x} = g G_x g^{-1}$ and the map $g \mapsto g \cdot x$ induces a bijection of G/G_x onto $G(x)$.

Definitions   1. A (linear) subspace U of a given linear G-space V is called an invariant (or G-) subspace, if

$$G \cdot U = \{g \cdot x : g \in G, x \in U\} \subseteq U.$$

2. A linear G-space V (or its corresponding representation of G on V) is said to be irreducible if $\{0\}$ and V are the only invariant subspaces.

3. A linear G-space V (or its corresponding representation of G on V) is called completely reducible if it can be expressed as the direct sum of *irreducible G-subspaces*.

4. The following equations define the induced linear G-space structures of two given linear G-spaces V and W.

(i) direct sum: $V \oplus W$ with $g \cdot (x, y) = (gx, gy)$.

(ii) dual space: V^* with $\langle x, g \cdot x' \rangle = \langle g^{-1} \cdot x, x' \rangle$. (Notice that the inverse in the above definition is needed to ensure that $\langle x, g_1 \cdot (g_2 \cdot x') \rangle = \langle x, (g_1 \cdot g_2) \cdot x' \rangle$ for all $x \in V$, $x' \in V^*$.)

(iii) tensor product: $V \otimes W$ with $g \cdot (x \otimes y) = g \cdot x \otimes g \cdot y$.

(iv) $\mathrm{Hom}(V, W)$: $A \in \mathrm{Hom}(V, W)$, $(g \cdot A)x = gA(g^{-1} \cdot x)$.

It follows from the above definition that the usual canonical isomorphisms such as

$$\mathrm{Hom}(V, W) \cong V^* \otimes W,$$

$$(V \otimes W)^* \cong V^* \otimes W^*,$$

$$U \otimes (V \oplus W) \cong (U \otimes V) \oplus (U \otimes W)$$

are automatically G-isomorphisms. Moreover, an element $A \in \mathrm{Hom}(V, W)$ is a fixed point if and only if it is a G-linear map. $g^{-1} \cdot A = A \Leftrightarrow$ for all $g \in G$, $g^{-1}A(gx) = A(x)$, i.e. $A(gx) = gA(x)$.

Of course, one may also define the induced G-space structure for the other linear algebra constructions such as $\Lambda^k(V)$, $S^k(V)$, etc., and again the canonical isomorphisms such as $V \otimes V \cong \Lambda^2(V) \oplus S^2(V)$ will also be G-isomorphisms.

Schur Lemma   *Let V, W be irreducible (linear) G-spaces and $A : V \to W$ be a G-linear map. Then A is either invertible or $A = 0$.*

Proof: Both $\ker A \subseteq V$ and $\mathrm{Im}\, A \subseteq W$ are clearly G-subspaces; it follows from the irreducibility assumption that

$$\ker A = \begin{cases} \{0\} \\ V, \end{cases} \quad \mathrm{Im}\, A = \begin{cases} \{0\} \\ W. \end{cases}$$

Therefore, the only possible combinations are exactly either (i) $\ker A = \{0\}$ and $\mathrm{Im}\, A = W$, i.e. A is invertible, or (ii) $\ker A = V$ and $\mathrm{Im}\, A = \{0\}$, i.e. $A = 0$. $\qquad\square$

In the special case of $V = W$ and the base field $\mathbb{C}$, one has the following refinement.

Special Form   *If V is an irreducible G-space over $\mathbb{C}$ and A is a G-linear self-map of V, then A is a scalar multiple, i.e. $A = \lambda_0 \cdot I$ for a suitable $\lambda_0 \in \mathbb{C}$.*

Proof: It is obvious that $A - \lambda I$ is also G-linear for any $\lambda \in \mathbb{C}$. Let λ_0 be an eigenvalue of A; this exists because $\mathbb{C}$ is algebraically closed. Then $A - \lambda_0 I$ is not invertible and hence must be zero, i.e. $A = \lambda_0 I$.   □

Corollary   *A complex irreducible representation of an Abelian group G is always one-dimensional.*

Proof: Let $\phi : G \to \mathrm{GL}(V)$ be a complex irreducible representation. Since G is commutative, $\phi(g) \cdot \phi(g_0) = \phi(g_0) \cdot \phi(g)$ for all $g_0, g \in G$. Hence, for each g, $\phi(g)$ is a G-linear self-map of V and therefore $\phi(g) = \lambda(g) \cdot I$ for a suitable $\lambda(g) \in \mathbb{C}$. g, however, is an arbitrary element of G, thus $\mathrm{Im}\,\phi = \{\phi(g) : g \in G\} \subset \mathbb{C}^* \cdot I$, the set of non-zero scalar multiples. Therefore any subspace of V is automatically G-invariant, and hence it can be irreducible only when $\dim V = 1$.   □

2. A Brief Overview

Before proceeding to the technical discussion of linear representation theory, let us pause a moment to reflect on some of the special features of linear transformation groups, to think about what are some of the natural problems that one might pursue and to have a brief overview of the fundamental results of such a theory.

Among all kinds of mathematical models, vector space structure is undoubtedly one of the most basic and most useful type; it is a kind of ideal combination of straightforward algebraic operations and simple, natural geometric intuitions. Correspondingly, linear transformation groups also inherit many advantageous and nice features. For example, they are conceptually rather elementary and concrete; they are easily accessible to algebraic computations; they can be readily organized by the canonical constructions of linear algebra, e.g., direct sum, tensor product, dual space, etc., and moreover, they also enjoy the beneficial help of geometric interpretation and imagination. Therefore, they are a kind of ideal material to serve as the "films" for taking "reconnaisance pictures".

The theory of representations of groups by linear transformation was created by G. Frobenius, here in Berlin during the years 1896–1903. His basic idea is that one should be able to obtain a rather wholesome understanding of the structure of a given group G by a systematic analysis of the totality of its "linear pictures". Next, let us try to formulate some natural problems along the above lines of thinking.

1. Problem on complete reducibility

If all representations of a given group G happen to be automatically completely reducible, then the study of linear G-spaces can easily be reduced to that of irreducible ones. Therefore, it is natural to ask "What type of groups have the property that all representations of such groups are automatically completely reducible"?

2. Problem on irreducibility criterion

How to determine whether a given representation is irreducible?

3. Problem on classification

How to classify irreducible representations of a given group G up to equivalence?

Finally, let us have a preview of some of the remarkable answers to the above basic problems obtained by G. Frobenius and I. Schur.

Theorem 1. *If G is a compact topological group, then any real (resp. complex) representation of G is automatically completely reducible.*

The key to the classification theory of linear representations of groups is the following invariant introduced by G. Frobenius.

Definition Let $\phi : G \to \mathrm{GL}(V)$ be a given complex representation of G. The complex-valued function

$$\chi_\phi : G \xrightarrow{\phi} \mathrm{GL}(V) \xrightarrow{\mathrm{tr}} \mathbb{C} : g \mapsto \mathrm{tr}\, \phi(g)$$

is called the character (function) of ϕ.

1. $\chi_\phi(g) = \operatorname{tr}\phi(g)$ is the sum with multiplicities of the eigenvalues of $\phi(g)$. Hence, it is quite obvious that equivalent representations have identical character functions, namely, the character function is an invariant of equivalent classes of representations.

2. If g_1, g_2 are conjugate in G, i.e. there is some $g \in G$ such that $g_1 = gg_2g^{-1}$, then

$$\chi_\phi(g_1) = \chi_\phi(gg_2g^{-1}) = \operatorname{tr}(\phi(g)\phi(g_2)\phi(g)^{-1}) = \operatorname{tr}\phi(g_2) = \chi_\phi(g_2).$$

Hence, the character function of an arbitrary representation ϕ of G has the special property of constancy on each conjugacy class of G.

3. If $\psi = \phi_1 \oplus \phi_2$, then it is easy to see that for all $g \in G$,

$$\chi_\psi(g) = \chi_{\phi_1}(g) + \chi_{\phi_2}(g),$$

namely, $\chi_\psi = \chi_{\phi_1} + \chi_{\phi_2}$ as functions.

The most remarkable result of Frobenius–Schur theory is the following classification theorem.

Theorem 2.   *If G is a compact topological group, then two representations ϕ and ψ are equivalent if and only if $\chi_\phi = \chi_\psi$ as functions.*

3. Compact Groups, Haar Integral and the Averaging Method

Let G be a finite group and V be a given linear G-space. Then, to each point $x \in V$, the center of mass of the orbit $G(x)$ is clearly a fixed point of V. Hence, the map

$$x \mapsto \bar{x} = \text{the center of mass of } G(x) = \frac{1}{|G|} \sum_{g \in G} g \cdot x$$

is a canonical projection of V onto V^G, the subspace of fixed points in V. In terms of a chosen coordinate system, the ith coordinate of $\bar{x}$ is simply the average value of the ith coordinate of $\{g \cdot x : g \in G\}$. We shall proceed to generalize the above useful method of producing fixed elements, namely, the averaging method, to the general setting of compact topological groups. Of course, the key step is to establish the correct meaning of the average value of a given continuous function $f : G \to \mathbb{R}$.

3.1. Haar integral of functions defined on compact groups

Let G be a given compact topological group and $C(G)$ be the linear space of all (real-valued) continuous functions of G equipped with the sup-norm topology. It is not difficult to show that every continuous function $f \in C(G)$ is automatically uniformly continuous, i.e. to any given $\delta > 0$, there exists a neighborhood U of the identity in G such that $xy^{-1} \in U \Rightarrow |f(x) - f(y)| < \delta$.

The translational transformation of $G \times G$ on G, namely,

$$T : (G \times G) \times G \to G : (g_1, g_2) \cdot x \mapsto g_1 x g_2^{-1}$$

naturally induces a continuous linear transformation of $G \times G$ on $C(G)$, namely,

$$[(g_1, g_2) \cdot f](x) = f(g_1^{-1} x g_2), \quad f \in C(G), \quad (g_1, g_2) \in G \times G.$$

Theorem 3. *There exists a unique G-projection $I : C(G) \to \mathbb{R}$ (the subspace of constant functions). $[I(f)$ is called the average value, or Haar integral, of f.]*

Proof: (a sketch) (i) Let A be a finite subset of $G \times G$ with multiplicities and $f \in C(G)$. Set $\Gamma(A, f)$ to be the center of mass of $A \cdot f$, namely,

$$\Gamma(A, f) = \frac{1}{|A|} \sum_{a \in A} m(a) \cdot a \cdot f,$$

where $m(a)$ is the multiplicity of a and $|A| = \sum m(a)$ is the total weight. For two finite subsets A, B of $G \times G$ with multiplicities, $A \cdot B$ is again a finite subset with multiplicities and it is easy to check that $\Gamma(A, \Gamma(B, f)) = \Gamma(A \cdot B, f)$.

(ii) Set $\Delta_f = \{\Gamma(A, f) : A$ is a finite subset with multiplicities of $G \times G\}$. For each $h \in C(G)$, set

$$\omega(h) = \max\{h(x) : x \in G\} - \min\{h(x) : x \in G\}.$$

Let $C(f)$ be the greatest lower bound of $\{\omega(h) : h \in \Delta_f\}$ and $\{h_n\}$ be a minimizing sequence, namely, $\omega(h_n) \to C(f)$ as $n \to \infty$. It is straightforward to check that Δ_f is a family of equicontinuous functions, namely, to any given $\delta > 0$, there exists a neighborhood U of the identity in G such that

$$(\forall h \in \Delta_f)xy^{-1} \in U \Rightarrow |h(x) - h(y)| < \delta.$$

Therefore, there exists a converging subsequence of $\{h_n\}$ and hence one may assume that $\{h_n\}$ is itself convergent to begin with.

(iii) Set $\bar{h} = \lim h_n$. Then it is clear that $\omega(\bar{h}) = C(f)$. Finally, one proves by contradiction that $\omega(\bar{h}) = C(f) = 0$! For otherwise, one can always choose a suitable finite subset $A \in G \times G$ such that

$$\omega(\Gamma(A, \bar{h})) < \omega(\bar{h}) = C(f).$$

Moreover, it is straightforward to check that $\Gamma(A, h_n)$ converges to $\Gamma(A, \bar{h})$ and $\lim \omega(\Gamma(A, h_n)) = \omega(\Gamma(A, \bar{h}))$. But all $\Gamma(A, h_n)$ are obviously also element of Δ_f, which contradicts the fact that $C(f)$ is the greatest lower bound for all $\omega(h)$. (We refer to L. S. Pontriagin's book *Topological Groups* for the details of the above proof due to von Neumann.) □

The above continuous $G \times G$-equivariant, linear functional $I : C(G) \to \mathbb{R}$ uniquely determines a $G \times G$-invariant measure of total measure 1 on G (called the Haar measure) such that $I(f) = \int_G f(g)dg$ for all $f \in C(G)$.

3.2. Existence of invariant inner (*resp. Hermitian*) product

As the first application of the averaging method, let us establish the following basic fact which includes Theorem 1 as an easy corollary.

Theorem 4. *Let V be a given real (resp. complex) linear G-space. If G is a compact topological group, then there exists a G-invariant inner (resp. Hermitian) product on V, namely*

$$(g \cdot x, g \cdot y) = (x, y) \quad \text{for all } x, y \in V, g \in G.$$

Proof: Let $\langle x, y \rangle$ be an arbitrary inner (resp. Hermitian) product on V. Set

$$(x, y) = \int_G \langle g \cdot x, g \cdot y \rangle dg.$$

It is straightforward to verify that (x, y) is again an inner (resp. Hermitian) product on V, and moreover

$$(a \cdot x, a \cdot y) = \int_G \langle ga \cdot x, ga \cdot y \rangle dg.$$

Letting $g' = ga$, $dg' = dg$, then

$$(a \cdot x, a \cdot y) = \int_G \langle g' \cdot x, g' \cdot y \rangle dg' = (x, y). \qquad \square$$

Definition   A real (resp. complex) linear G-space with an invariant inner (resp. Hermitian) product is called an orthogonal (resp. unitary) G-space, and the corresponding representation is called an orthogonal (resp. unitary) representation.

In an orthogonal (resp. unitary) G-space V, the perpendicular subspace to an *invariant* subspace is automatically also an invariant subspace.

Proof of Theorem 1:   By Theorem 4, one may equip V with an invariant inner (resp. Hermitian) product. Let U by a positive dimensional irreducible sub-G-space of V and $U^\perp$ be its perpendicular subspace. Then $V = U \oplus U^\perp$ is a decomposition of V into the direct sum of sub-G-spaces, $\dim U^\perp < \dim V$. From here, the proof of Theorem 1 follows by a simple induction on $\dim V$. $\qquad \square$

Exercises   1. Let $O(n) \subset \mathrm{GL}(n, \mathbb{R})$ (resp. $U(n) \subset \mathrm{GL}(n, \mathbb{C})$) be the subgroup of orthogonal (resp. unitary) matrices. Show that they are compact.

2. Let $G \subset \mathrm{GL}(n, \mathbb{R})$ (resp. $\mathrm{GL}(n, \mathbb{C})$) be a compact subgroup. Show that there exists a suitable element A in $\mathrm{GL}(n, \mathbb{R})$ (resp. $\mathrm{GL}(n, \mathbb{C})$) such that

$$AGA^{-1} \subset O(n) \quad (\text{resp. } U(n)).$$

3. Show that $O(n)$ (resp. $U(n)$) is a *maximal* compact subgroup of $\mathrm{GL}(n, \mathbb{R})$ (resp. $\mathrm{GL}(n, \mathbb{C})$) and any two maximal compact subgroups of $\mathrm{GL}(n, \mathbb{R})$ (resp. $\mathrm{GL}(n, \mathbb{C})$) must be mutually conjugate.

4. Let ϕ, ψ be two complex representations of a compact group G. Then $\chi_{\phi \otimes \psi}(g) = \chi_\phi(g) \cdot \chi_\psi(g)$ for all $g \in G$. (Thanks to Theorem 4, $\phi(g)$ and $\psi(g)$ are always diagonalizable.)

4. Frobenius–Schur Orthogonality and the Character Theory

Now let us apply the averaging method to analyze the deep implications of the Schur lemma.

Case 1: Let $\phi : G \to \mathrm{GL}(V)$, $\psi : G \to \mathrm{GL}(W)$ be two *non-equivalent, irreducible* complex representations of a compact group G. Then it follows from the Schur lemma that

$$\mathrm{Hom}_G(V, W) = \mathrm{Hom}(V, W)^G = \{0\}.$$

(Recall that X^G denotes the fixed point set of a G-space X.) Therefore, it follows from the averaging method that for all $A \in \mathrm{Hom}(V, W)$

$$\int_G g \cdot A \, dg = \int_G \psi(g) \cdot A \cdot \phi(g)^{-1} dg = 0,$$

because $\int_G g \cdot A \, dg$ is the center of mass of $G(A)$ and, of course, it is always a fixed point!

By Theorem 4, one may equip both V and W with invariant Hermitian products and compute the above powerful equation in its matrix form with respect to chosen orthonormal bases in V and W. Let E_{ik} be the linear map which maps the kth base vector of V to the ith base vector of W and all the other base vectors of V to zero.

Since the above equation is linear with respect to the parameter A and $\{E_{ik} : 1 \leq i \leq \dim W, 1 \leq k \leq \dim V\}$ already forms a basis of $\mathrm{Hom}(V, W)$, one needs only to compute the special cases of $A = E_{ik}$. Set

$$\phi(g) = (\phi_{kl}(g)), \quad \psi(g) = (\psi_{ij}(g)).$$

$(\phi_{kl}(g), \psi_{ij}(g) \in C(G)$ are called representation functions.) One has

$$0 = \int_G g \cdot E_{ab} dg = \int_G (\psi_{ij}(g)) \cdot E_{ab} \cdot (\bar{\phi}_{kl}(g))^t dg$$

$$= \int_G (\psi_{ia}(g) \cdot \bar{\phi}_{kb}(g)) dg.$$

Hence

$$\int_G \psi_{ia}(g) \cdot \bar{\phi}_{kb}(g) dg = 0,$$

for $1 \leq i, a \leq \dim W$, $1 \leq k, b \leq \dim V$.

Case 2: The special form of the Schur lemma asserts that

$$\text{Hom}_G(V, V) = \text{Hom}(V, V)^G = \{\lambda \cdot I : \lambda \in C^*\}.$$

Hence, it again follows from the averaging method that

$$\int_G g \cdot B dg = \int_G \phi(g) \cdot B \cdot \phi(g)^{-1} dg = \lambda_B \cdot I,$$

where λ_B is a yet-to-be-determined complex number solely depending on B. Exploiting the linearity and the conjugate invariance of the trace, one has

$$\lambda_B \cdot \dim V = \text{tr} \, \lambda_B \cdot I = \text{tr} \int_G \phi(g) \cdot B \cdot \phi(g)^{-1} dg$$

$$= \int_G \text{tr}(\phi(g) \cdot B \cdot \phi(g)^{-1}) dg = \int_G \text{tr} \, B dg$$

$$= \text{tr} \, B$$

which determines the value of λ_B, namely

$$\lambda_B = \frac{1}{\dim V} \text{tr} \, B.$$

From here, the same computation as that of Case 1 will yield the following set of equations, namely

$$\int_G \phi_{ij}(g) \cdot \bar{\phi}_{kl}(g) dg = \frac{1}{\dim V} \delta_{ik} \delta_{jl},$$

for $1 \leq i, j, k, l \leq \dim V$.

Summarizing the above fundamental results, we state them as the following theorem:

Theorem 5.   *Let $\phi(g) = (\phi_{kl}(g))$, $\psi(g) = (\psi_{ij}(g))$ be two non-equivalent irreducible unitary representations of a compact group G. Then*

$$\int_G \psi_{ij}(g) \cdot \bar{\phi}_{kl}(g) dg = 0,$$

$$\int_G \phi_{ij}(g) \cdot \bar{\phi}_{kl}(g) dg = \frac{1}{\dim V} \delta_{ik} \delta_{jl}.$$

Corollary 1.

$$\int_G \chi_\phi(g) \cdot \bar{\chi}_\phi(g) dg = 1,$$

$$\int_G \chi_\psi(g) \cdot \bar{\chi}_\phi(g) dg = 0.$$

Proof: By Definition,

$$\chi_\phi(g) = \sum_{k=1}^{\dim \phi} \phi_{kk}(g), \quad \chi_\psi(g) = \sum_{i=1}^{\dim \psi} \psi_{ii}(g).$$

Hence, the above statements follow from a direct application of Theorem 5. □

Let $\hat{G}$ be the set of *equivalence classes of complex irreducible* representations of a given compact group G. It follows from Theorem 1 that every complex representation ρ of G can be expressed as the direct sum of irreducible ones, namely

$$\rho = \sum_{\phi \in \hat{G}} \oplus m(\rho; \phi) \cdot \phi,$$

where $m(\rho; \phi)$ is the multiplicity of irreducible representations of the equivalence class ϕ in the decomposition of ρ.

Corollary 2.

$$m(\rho; \phi) = \int_G \chi_\rho(g) \cdot \bar{\chi}_\phi(g) dg,$$

$$\int_G \chi_{\rho_1}(g) \cdot \bar{\chi}_{\rho_2}(g) dg = \sum_{\phi \in \hat{G}} m(\rho_1; \phi) \cdot m(\rho_2; \phi).$$

Proof:

$$\chi_\rho(g) = \sum_{\phi \in \hat{G}} m(\rho; \phi) \chi_\phi(g).$$

Hence, the above two equations follow immediately from Corollary 1. □

Theorem 6 follows easily from Corollary 2; we restate it in the following slightly more precise form.

Theorem 6. *Two complex representations ρ, ψ of a compact group G are equivalent if and only if $\chi_\rho = \chi_\psi$ (as functions). A complex representation ρ is irreducible if and only if*

$$\int_G \chi_\rho(g) \cdot \bar{\chi}_\rho(g) dg = 1.$$

Proof: It is obvious that $\rho \sim \psi \Rightarrow \chi_\rho(g) = \chi_\psi(g)$, namely

$$\chi_\rho(g) = \operatorname{tr} \rho(g) = \operatorname{tr}(A\rho(g)A^{-1}) = \operatorname{tr} \psi(g) = \chi_\psi(g).$$

Conversely, $\chi_\rho = \chi_\psi$ (as functions) implies that

$$m(\rho; \phi) = \int_G \chi_\rho(g) \cdot \bar{\chi}_\phi(g) dg = \int_G \chi_\psi(g) \cdot \bar{\chi}_\phi(g) dg = m(\psi; \phi),$$

for all $\phi \in \hat{G}$. Hence $\rho \sim \psi$. Finally,

$$\int_G \chi_\rho(g) \cdot \bar{\chi}_\rho(g) dg = \sum_{\phi \in \hat{G}} m(\rho; \phi)^2 = 1$$

simply means that there is exactly one $m(\rho; \phi) = 1$ and the rest of them are all zero! Hence ρ is itself irreducible. □

A Classical Example $G = S^1 = \{e^{i\theta}; 0 \leq \theta < 2\pi\}$. To each integer $n \in \mathbb{Z}$, there is a one-dimensional complex representation

$$\phi : S^1 \times \mathbb{C}^1 \to \mathbb{C}^1 : e^{i\theta} \cdot z = e^{in\theta} z,$$

or equivalently,

$$S^1 \overset{\phi}{\to} U(1) = S^1, \quad e^{i\theta} \mapsto e^{in\theta}.$$

In this special case, the above results specialize into the well-known facts in the Fourier series, namely, that $\{e^{in\theta} : n \in \mathbb{Z}\}$ forms an orthonormal basis of $L_2(S^1)$.

Exercises 1. Use the completeness of $\{e^{in\theta} : n \in \mathbb{Z}\}$ in $L_2(S^1)$ to show that the above collection of representations of S^1 already forms a complete set of representatives of $\hat{S}^1$.

2. Generalize the above result of S^1 to products of several copies of S^1, namely, the torus group of rank k:

$$T^k = S^1 \times S^1 \times \cdots \times S^1 \quad (k \text{ copies}).$$

Hint: Exhibit a collection of explicit irreducible complex representations of T^k (notice that they must all be one-dimensional!) and then apply the above theory on representation functions to check whether you have already obtained a complete collection of representatives of $\hat{T}^k$.

5. Classification of Irreducible Complex Representations of S^3

Among all compact connected non-Abelian topological groups, the multiplicative group of unit quaternions, S^3, is certainly the simplest one and is also one of the most basic. As a preliminary application of the character theory of Sec. 4, let us work out the classification problem of irreducible complex representations of S^3 as follows.

Let $\mathbf{H} = \{a + jb : a, b \in \mathbb{C}\}$ be the skew field of quaternions and $S^3 = \{a + jb : a\bar{a} + b\bar{b} = 1\}$ be the multiplicative group of unit quaternions. We shall consider $\mathbf{H}$ as a *right* $\mathbb{C}$-module and let S^3 act on $\mathbf{H}$ via left multiplications. (In this setting, the associative law of $\mathbf{H}$ shows that the S^3-action on $\mathbf{H}$ is indeed $\mathbb{C}$-linear.) Choose $\{1, j\}$ as the $\mathbb{C}$-basis of $\mathbf{H}$. Then the above S^3-action gives a two-dimensional representation:

$$\phi_1 : S^3 \to U(2) : \phi(a + jb) = \begin{pmatrix} a & -\bar{b} \\ b & \bar{a} \end{pmatrix}.$$

(Note $(a + jb) \cdot j = aj + jbj = -\bar{b} + j\bar{a}$.) In fact, the above map is an isomorphism of S^3 onto $SU(2)$, the subgroup of $U(2)$ with determinant 1.

We can interpret the above matrix as the following linear substitution:

$$\begin{pmatrix} z_1 \\ z_2 \end{pmatrix} \mapsto \begin{pmatrix} a & -\bar{b} \\ b & \bar{a} \end{pmatrix} \begin{pmatrix} z_1 \\ z_2 \end{pmatrix},$$

and hence it induces a linear transformation on the space of polynomials

$$\mathbb{C}[z_1, z_2] = \sum_{k=0}^{\infty} \oplus V_k,$$

where V_k is the subspace of homogeneous polynomials of degree k. Each V_k is clearly an *invariant* subspace of the above $SU(2)$-action and it is of dimension $k + 1$. Thus, the restricted $SU(2)$-action on V_k produces a complex representation of dimension $k + 1$ for each $k = 0, 1, 2, \dots$.

Theorem 7. *Let ϕ_k be the above complex representation of S^3 on V_k. Then each ϕ_k is irreducible and they form a complete set of representatives of $\hat{G}$ for $G = S^3$.*

Proof: (i) Since all character functions are automatically constant along each conjugacy class of G, a good understanding of the orbital geometry of the adjoint transformation, namely

$$\text{Ad} : G \times G \to G : (g, x) \mapsto gxg^{-1}$$

will be very helpful in the actual computation of integration of such functions over G.

Consider $\mathbf{H}$ as a four-dimensional real vector space with inner product and let S^3 act on it as follows:

$$S^3 \times \mathbf{H} \to \mathbf{H} : (g, x) \mapsto g \cdot x \cdot g^{-1} \quad \text{(quaternion multiplication)}.$$

It leaves the line of real numbers pointwise fixed and it preserves the norm. Hence it is an orthogonal representation of the form $1 \oplus \psi$ where 1 denotes the trivial representation acting on the real line and ψ is the restriction of the above S^3-action on the $\mathbb{R}^3$ of pure quaternions. Therefore

$$\psi : S^3 \to \text{SO}(3),$$

and it is easy to see $\ker \psi = \{\pm 1\}$. Since $\dim \text{SO}(3) = 3 = \dim S^3/\{\pm 1\}$, it is clear that ψ is an epimorphism.

The adjoint transformation of S^3 is exactly the restriction of the above S^3-action of $\mathbf{H}$ to the unit sphere $S^3(1)$. Hence, every conjugacy class of S^3 intersects the subgroup S^1 of unit complexes perpendicularly at conjugate points and the conjugacy class of $e^{\pm i\theta}$ is the "latitude" two-sphere passing through $e^{\pm i\theta}$, which is intrinsically a two-sphere of radius $\sin \theta$. (See Fig. 1.)

(ii) Since

$$\phi_1(e^{i\theta}) \cdot z_1 = e^{i\theta} \cdot z_1, \quad \phi_1(e^{i\theta}) \cdot z_2 = e^{-i\theta} \cdot z_2,$$

it is easy to see that

$$z_1^k, z_1^{k-1}z_2, \ldots, z_1^{k-j}z_2^j, \ldots, z_2^k$$

are eigenvectors of $\phi_k(e^{i\theta})$ with

$$e^{ik\theta}, e^{i(k-2)\theta}, \ldots, e^{i(k-2j)\theta}, \ldots, e^{-ik\theta}$$

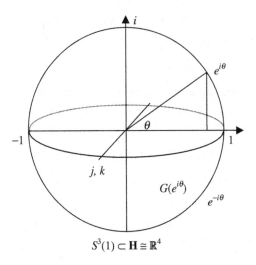

Fig. 1.

as their respective eigenvalues. Hence

$$\chi_k(e^{i\theta}) = \chi_{\phi_k}(e^{i\theta}) = \frac{e^{i(k+1)\theta} - e^{-i(k+1)\theta}}{e^{i\theta} - e^{-i\theta}}.$$

(iii) The total volume of $S^3(1)$ is $2\pi^2$. Let $d\sigma$ be the volume element of the Riemannian manifold $S^3(1)$ and dg be the normalized Haar measure. Then $dg = \frac{1}{2\pi^2}d\sigma$.

(iv) The nice orbit geometry of (i) enables us to exploit the orbital constancy property of the character functions to simplify the integrations over S^3, namely

$$\int_G \chi_k(g) \cdot \bar{\chi}_k(g) dg$$

$$= \frac{1}{2\pi^2} \int_{S^3(1)} \chi_k(g) \cdot \bar{\chi}_k(g) d\sigma$$

$$= \frac{1}{2\pi^2} \int_0^\pi \chi_k(e^{i\theta}) \cdot \bar{\chi}_k(e^{i\theta}) \cdot 4\pi \sin^2 \theta d\theta$$

$$= \frac{1}{4\pi} \int_0^{2\pi} \chi_k(e^{i\theta}) \cdot \bar{\chi}_k(e^{i\theta}) \cdot |e^{i\theta} - e^{-i\theta}|^2 d\theta$$

$$= \frac{1}{4\pi} \int_0^{2\pi} |e^{i(k+1)\theta} - e^{-i(k+1)\theta}|^2 d\theta$$

$$= 1.$$

Hence, by Theorem 6, ϕ_k is irreducible!

(v) Finally, we shall prove the *completeness* of $\{\phi_k\}$ by contradiction. Suppose ψ is an irreducible complex representation of dimension $k+1$ but it is non-equivalent to ϕ_k. Then

$$0 = \int_G \chi_\psi(g) \cdot \bar{\chi}_l(g) dg$$

$$= \frac{1}{4\pi} \int_0^{2\pi} \chi_\psi(e^{i\theta}) \cdot \bar{\chi}_l(e^{i\theta}) \cdot |e^{i\theta} - e^{-i\theta}|^2 d\theta$$

$$= \frac{1}{4\pi} \int_0^{2\pi} \chi_\psi(e^{i\theta}) \cdot (e^{i\theta} - e^{-i\theta}) \cdot \overline{(e^{i(l+1)\theta} - e^{-i(l+1)\theta})} d\theta,$$

for all non-negative integers l because ψ is not equivalent to any ϕ_l. (Note that $\dim \phi_l \neq \dim \psi$ if $l \neq k$.)

Observe that $\chi_\psi(e^{i\theta}) = \chi_\psi(e^{-i\theta})$ and hence it is an even function of θ. Therefore $\chi_\psi(e^{i\theta}) \cdot (e^{i\theta} - e^{-i\theta})$ is a non-zero odd function of θ which, by the above equation, is orthogonal to all $\{(e^{i(l+1)\theta} - e^{-i(l+1)\theta}) : l = 0, 1, 2, \ldots\}$. This is a contradiction to the well-known fact that they already form a basis of the subspace of odd L_2-functions of S^1. Hence, such an irreducible representation ψ cannot possibly exist. This proves that the family $\{\phi_k\}$ already constitutes a complete representatives of $\hat{G}$ for $G = S^3 \cong \text{SU}(2)$. $\square$

Exercise Using the fact $\text{SO}(3) \cong S^3/\{\pm 1\}$, every irreducible representation $\phi : \text{SO}(3) \to \text{GL}(V)$ can always be "pulled back" to an irreducible representation with $\ker \phi \supset \{\pm 1\}$, namely

$$\tilde{\phi} : S^3 \xrightarrow{\pi} \text{SO}(3) \xrightarrow{\phi} \text{GL}(V), \quad \tilde{\phi} = \phi \circ \pi.$$

Conversely, every irreducible representation of S^3 whose ker contains $\{\pm 1\}$ can be considered as such a pull-back. Use the above relation to classify complex irreducible representations of $\text{SO}(3)$.

6. $L_2(G)$ and Concluding Remarks

The results of Frobenius–Schur theory clearly indicate that $L_2(G)$ should be a proper setting for further development of representation theory of compact groups. Therefore, we shall conclude our rather brief discussion on representation theory by mentioning some pertinent results along this line.

1. Theorem 5 proves that

$$\left\{ \sqrt{\dim \phi} \cdot \phi_{ij} : \phi \in \hat{G}, 1 \leq i, j \leq \dim \phi \right\}$$

is a natural collection of orthonormal vectors in $L_2(G)$ and

$$\{\chi_\phi : \phi \in \hat{G}\}$$

is a natural collection of orthonormal vectors in $L_2(G)^{\mathrm{Ad}} \cong L_2(G/\mathrm{Ad})$. Of course, it would be nice if they actually formed orthonormal bases of $L_2(G)$ and $L_2(G/\mathrm{Ad})$ respectively. Indeed, this is exactly the assertion of the Peter–Weyl theorem. We refer to Pontriagin's book *Topological Groups* for a proof of this basic theorem.

2. Let $G = G_1 \times G_2$ and ϕ_1, ϕ_2 be complex irreducible representations of G_1, G_2 on V_1, V_2 respectively. Then G has a natural induced action on $V_1 \otimes V_2$, namely

$$(g_1, g_2) \cdot (x_1 \otimes x_2) = \phi_1(g_1)x_1 \otimes \phi_2(g_2)x_2.$$

We shall call it the *outer* tensor product of ϕ_1 and ϕ_2 and will be denoted by $\phi_1 \hat{\otimes} \phi_2$. It is easy to check that

$$\chi_{\phi_1 \hat{\otimes} \phi_2}(g_1, g_2) = \chi_{\phi_1}(g_1) \cdot \chi_{\phi_2}(g_2).$$

Hence

$$\int_{G_1 \times G_2} |\chi_{\phi_1 \hat{\otimes} \phi_2}(g_1, g_2)|^2 dg = \int_{G_1 \times G_2} |\chi_{\phi_1}(g_1)|^2 \cdot |\chi_{\phi_2}(g_2)|^2 dg_1 \cdot dg_2$$

$$= \int_{G_1} |\chi_{\phi_1}(g_1)|^2 dg_1 \cdot \int_{G_2} |\chi_{\phi_2}(g_2)|^2 dg_2$$

$$= 1 \cdot 1 = 1.$$

This proves that the *outer tensor product* of two complex irreducible representations of G_1, G_2 is always an irreducible complex representation of $G_1 \times G_2$.

Caution! Notice the difference between the outer tensor product and the previous tensor product defined for two representations of the same group. In fact, if ϕ and ψ are two representations of the same group G, then one has the following commutative diagram of homomorphisms:

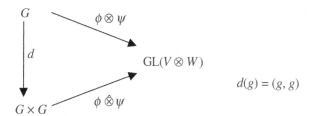

$$d(g) = (g, g)$$

3. We just showed that

$$\phi \in \hat{G}_1, \psi \in \hat{G}_2 \Rightarrow \phi \hat{\otimes} \psi \in \widehat{G_1 \times G_2}.$$

In fact, $\widehat{G_1 \times G_2} = \{\phi \hat{\otimes} \psi : \phi \in \hat{G}_1, \psi \in \hat{G}_2\}$. The proof of this fact is as follows:

$\{\chi_\phi(g_1) : \phi \in \hat{G}_1\}$ forms an orthonormal basis of $L_2(G_1/\mathrm{Ad})$,

$\{\chi_\psi(g_2) : \psi \in \hat{G}_2\}$ forms an orthonormal basis of $L_2(G_2/\mathrm{Ad})$,

$\dfrac{G_1 \times G_2}{\mathrm{Ad}} \cong (G_1/\mathrm{Ad}) \times (G_2/\mathrm{Ad})$ with product measure.

Hence, it follows from the well-known general fact that

$$\{\chi_\phi(g_1) \cdot \chi_\psi(g_2) : \phi \in \hat{G}_1, \psi \in \hat{G}_2\}$$

also forms an orthonormal basis of $L_2(G_1/\mathrm{AD} \times G_2/\mathrm{Ad}) \cong L_2(\frac{G_1 \times G_2}{\mathrm{Ad}})$. This proves that

$$\{\phi \hat{\otimes} \psi : \phi \in \hat{G}_1, \psi \in \hat{G}_2\} = \widehat{G_1 \times G_2}.$$

4. The $G \times G$-action of G given by $(g_1, g_2) \cdot x = g_1 x g_2^{-1}$ induces a $G \times G$-action on $L_2(G)$. It is not difficult to see that $L_2(G)$ decomposes into the direct sum of the following irreducible $G \times G$-subspaces, namely, for each $\phi \in \hat{G}$, one has the subspace spanned by $\{\phi_{ij} : 1 \le i, j \le \dim \phi\}$.

5. In the special case that G is a finite group, one has

(i) $\dim L_2(G) = |G|$ (the order of G),

(ii) $|\hat{G}| = \dim L_2(G/\mathrm{Ad}) = |G/\mathrm{Ad}|$, i.e. the number of distinct complex irreducible representations is equal to the number of conjugacy classes.

(iii) The decomposition of $L_2(G)$ yields the following interesting equation:

$$|G| = \sum_{\phi \in \hat{G}} (\dim \phi)^2.$$

Exercises   1. Find the relationship between the irreducible representation ϕ of G and the above irreducible representation of $G \times G$ on the subspace in $L_2(G)$ spanned by $\{\phi_{ij}(g) : 1 \le i, j \le \dim \phi\}$.

2. Apply the character theory to classify complex irreducible representations of the polyhedral groups, i.e., the symmetry groups of regular solids.

Lecture 2

Lie Groups and Lie Algebras

A Lie group G is, by definition, a *differentiable group*; it consists of a group structure and a manifold structure such that the multiplication map and the inversion map are differentiable. One might say that the vector space structure is a natural focal point of various branches of mathematics at its elementary level. The Lie group structure is another natural focal point at its higher ground. Intuitively speaking, the differentiability of the group structure should provide a way to "linearize" the group structure at the "infinitesimal level" and the "linear object" so obtained should be a useful invariant in analyzing the original Lie group structure. This decisive step was accomplished by S. Lie in the late nineteenth century. The linear object he obtained was originally called the "infinitesimal group" by himself and was later renamed to "Lie algebra" by H. Weyl.

Methodologically, it is rather interesting to note that the scheme that we are going to use is exactly the dual of the scheme that one uses in representation theory, namely, instead of taking "reconnaissance pictures" for analyzing a given structure, one sends "spies" into the structure to probe it directly! In fact, even the analytical tools that one uses in the above two approaches are also dual to each other, namely, integration and averaging for the former, differentiation and existence and uniqueness of solutions of ordinary differential equations for the latter.

1. One-Parameter Subgroups and Lie Algebras

Suppose one is planning to send a probing agent to study the structure of a given Lie group G. Of course, the success of the whole program depends on the selection of an effective agent. It is a mere common sense that such an agent should be both simple and flexible so that is can easily submerge itself into almost everywhere in G without disturbing the structure of G. A moment of reflection along this line will lead us to call for the help of our wonderful old friend the additive group of real numbers, which is the simplest Lie group.

Definition   A differentiable homomorphism of $(\mathbb{R}, +)$ into a given Lie group G, $\phi : \mathbb{R} \to G$, is called a one-parameter subgroup of G.

The initial velocity of ϕ, $\frac{d\phi}{dt}|_{t=0}$, is an element in the tangent space of G at the identity e, $T_e G$. One of the first natural basic questions is, of course, the following existence and uniqueness problem.

Uniqueness   Is a one-parameter subgroup $\phi : \mathbb{R} \to G$ uniquely determined by its initial velocity vector?

Existence   Can every element of $T_e G$ be realized as the initial velocity vector of a one-parameter subgroup of G?

The following analysis will naturally lead to an affirmative answer of the above problems in both the uniqueness and the existence.

Let $\phi : \mathbb{R} \to G$ be a given one-parameter subgroup. Then one may combine it with the right (resp. left) translation to obtain a *left-* (resp. *right-*) *invariant* $\mathbb{R}$-action on G, namely

$$\Phi : \mathbb{R} \times G \to G : \Phi(t, x) = x \cdot \phi(t) \quad (\text{resp. } \phi(t) \cdot x).$$

The left- (resp. right-) invariance means that

$$\Phi(t, a \cdot x) = a \cdot \Phi(t, x) \quad (\text{resp. } \Phi(t, x \cdot a) = \Phi(t, x) \cdot a),$$

which, of course, follows from the associativity. Following the usual convention, we shall always use the right translation so that the corresponding $\mathbb{R}$-action is left-invariant.

The velocity vectors of the above $\mathbb{R}$-action constitute a left-invariant vector field $\tilde{X}$ on G, i.e. for every left translation $l_a : G \to G : l_a(x) = a \cdot x$, $dl_a(\tilde{X}_x) = \tilde{X}_{ax}$. Moreover, it is quite obvious that a left-invariant vector field $\tilde{X}$ on G is uniquely determined by its value at the identity, namely, the map

$$\{\text{left invariant vector fields } \tilde{X}\} \to \{X = \tilde{X}_e \in T_e G\}$$

is a bijection. For all $x \in G$, $\tilde{X}_x = dl_x(\tilde{X}_e)$.

Let $\tilde{X}$ be a given left-invariant vector field on G. Then applying the usual existence and uniqueness theorem on systems of first-order ODE, passing through every point $x \in G$ there exists a unique integral curve whose velocity vectors all belong to $\tilde{X}$. Let $\phi : \mathbb{R} \to G$ be the unique integral curve of $\tilde{X}$ with $\phi(0) = e$. It follows from the left-invariance of $\tilde{X}$ that $l_a \circ \phi : \mathbb{R} \to G : t \mapsto a \cdot \phi(t)$ is the unique integral curve of $\tilde{X}$ with a as its initial point. Hence in particular

$$\phi(s) \cdot \phi(t) = \phi(s + t),$$

namely, $\phi : \mathbb{R} \to G$ is, in fact, a one-parameter subgroup of G.

Summarizing the above discussions, one has the following natural bijections between the following four types of related objects:

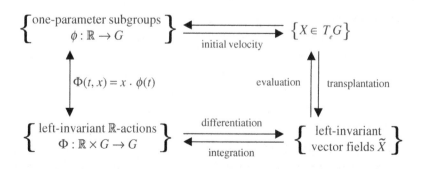

The only unmarked arrow "$\leftarrow$" is the composition of transplantation, integration and restriction to the integral curve with e as its initial point.

As one might notice, our old friend $\mathbb{R}$ skillfully uses four different "passports" in carrying out his mission successfully. Furthermore, in analyzing the above beautiful final report of his mission, one finds that it inherits a vector space structure from T_eG and a bracket operation from that of left invariant vector fields, because the bracket $[\tilde{X}, \tilde{Y}]$ of two left-invariant vector fields is clearly also left-invariant. Therefore the final result one obtains is a vector space T_eG with an additional bilinear anti-commutative bracket operation satisfying the usual Jacobi identity:

$$[[X, Y], Z] + [[Y, Z], X] + [[Z, X], Y] \equiv 0.$$

This is exactly the linearized object of a given Lie group structure which S. Lie called it the infinitesimal group of G, but nowadays, we call it the Lie algebra of G, denoted by $\mathfrak{G}$.

As it turns out, the above type of "Lie algebra" structure is not only important for the study of Lie groups, but it is also a powerful tool in many other branches of mathematics. Therefore it certainly deserves an independent standing and an independent theory for its own sake. Actually, this was exactly the reason why H. Weyl proposed to change the name "infinitesimal group" to the more independent-looking name "Lie algebra".

Definition   A Lie algebra over a field F is a vector space, which may be infinite dimensional, together with a bilinear, anti-commutative binary operation satisfying the Jacobi identity.

In fact, it is possible to organize the totality of all one-parameter subgroups of a given Lie group G into a single map of $\mathfrak{G}$ into G.

Definition   For each $X \in \mathfrak{G}$, set $\operatorname{Exp} X = \phi_X(1)$, where ϕ_X is the unique one-parameter subgroup of G with X as its initial velocity vector. The map so defined

$$\operatorname{Exp} : \mathfrak{G} \to G : X \mapsto \phi_X(1)$$

is called the exponential map of G.

Observe that $\mu_c : \mathbb{R} \to \mathbb{R} : \mu_c(t) = c \cdot t$ is obviously a Lie homomorphism. Hence, $\phi_X \circ \mu_c$ is again a one-parameter subgroup of G and it follows from the chain rule of differentiation that $\phi_X \circ \mu_c = \phi_{cX}$. Therefore for all $t \in \mathbb{R}$,

$$\operatorname{Exp} tX = \phi_{tX}(1) = \phi_X(t).$$

To put the above organization into perspective, one has the following commutative diagram. For each $X \in \mathfrak{G}$, one has

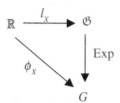

where l_X is the unique linear map $\mathbb{R} \to \mathfrak{G}$ with $l_X(1) = X$ and ϕ_X is the unique one-parameter subgroup with X as initial velocity. Thus Exp: $\mathfrak{G} \to G$ is actually the "universal map" for all one-parameter subgroups of G.

Let f be an arbitrarily given smooth function on G. Then $f_a^X(t) = f(a \cdot \mathrm{Exp}\, tX) \in C^\infty(\mathbb{R})$ is the pull-back of f, namely

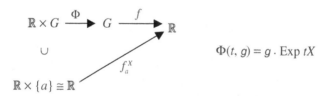

and moreover, $Df_a^X(t) = Xf(a \cdot \mathrm{Exp}\, tX)$. Therefore, the usual Taylor's formula with remainder, applied to $f_a^X(t)$, can be translated as follows:

$$f(a \cdot \mathrm{Exp}\, t_0 X) = f_a^X(t_0)$$

$$= f_a^X(0) + Df_a^X(0)t_0 + \frac{1}{2}D^2 f_a^X(0)t_0^2 + \cdots$$

$$+ \frac{1}{k!}D^k f_a^X(0)t_0^k + \frac{1}{(k+1)!}D^{k+1}f_a^X(\theta)t_0^{k+1}$$

$$= f(a) + Xf(a)t_0 + \cdots + \frac{1}{k!}X^k f(a)t_0^k$$

$$+ \frac{t_0^{k+1}}{(k+1)!}X^{k+1}f(a \cdot \mathrm{Exp}\, \theta X),$$

where θ is a suitable number between 0 and t_0.

Based upon the above Taylor's formula for smooth functions on G, it is straightforward to show that

$$\text{Exp}\, sX \cdot \text{Exp}\, tY \equiv \text{Exp}(sX + tY) \quad (\text{mod second-order terms})$$

$$\text{Exp}\, sX \cdot \text{Exp}\, tY \cdot \text{Exp}(-sX) \cdot \text{Exp}(-tY) \equiv \text{Exp}\, st[X, Y]$$

$$(\text{mod third-order terms}).$$

To be more precise, the above "$\equiv$" means the coordinates of both sides are equal modulo second- (resp. third-) order infinitesimals. Hence, the vector space structure of $\mathfrak{G}$ approximates the group operation of G up to the first order of infinitesimal, and the bracket operation of $\mathfrak{G}$ records the leading term of the *non-commutativity* of G which is a second-order infinitesimal.

Summarizing the above discussions, we state the results obtained so far as the following theorem.

Theorem 1.   (i) *To each tangent vector $X \in T_e G$ at the identity e, there exists a unique one-parameter subgroup $\phi_X : \mathbb{R} \to G$ with X as its initial velocity.*

(ii) *There exist canonical bijections between the following four sets of objects associated with a given Lie group $G : T_e G = \{\text{tangent vectors at } e\}$, $\{\text{one-parameter subgroups}\}$, $\{\text{left-invariant } \mathbb{R}\text{-actions}\}$, $\{\text{left-invariant vector fields}\}$.*

(iii) *The vector space $\mathfrak{G} = T_e G$ has an additional bracket operation (obtained from its canonical bijection with the space of left-invariant vector fields of G) which is bilinear, anti-commutative and satisfying the Jacobi identity. It is called the Lie algebra of G.*

(iv) *The totality of all one-parameter subgroups of G can be organized into an exponential map $\text{Exp} : \mathfrak{G} \to G$, such that $\phi_X(t) = \text{Exp}\, tX$.*

(v) *For each $f \in C^\infty(G)$, $a \in G$, $t_0 \in \mathbb{R}$, one has the following Taylor expansion with remainder:*

$$f(a \cdot \text{Exp}\, t_0 X) = f(a) + t_0 X f(a) + \cdots + \frac{t_0^k}{k!} X^k f(a)$$

$$+ \frac{t_0^{k+1}}{(k+1)!} X^{k+1} f(a \cdot \text{Exp}\, \theta X).$$

(vi) *To each Lie homomorphism $h : G_1 \to G_2$, its differential at e, $dh_e : \mathfrak{G}_1 \to \mathfrak{G}_2$ is a Lie algebra homomorphism.*

Examples 1. In the case of $\mathrm{GL}(n, \mathbb{R})$ (resp. $\mathrm{GL}(n, \mathbb{C})$), the following exponential power series of matrices

$$\mathrm{Exp}\, A = I + A + \frac{1}{2}A^2 + \cdots + \frac{1}{k!}A^k + \cdots$$

defines a map $\mathrm{Exp} : M_{n,n}(\mathbb{R}) \to \mathrm{GL}(n, \mathbb{R})$ (resp. $M_{n,n}(\mathbb{C}) \to \mathrm{GL}(n, \mathbb{C})$). It is well-known that $\phi_A(t) = \mathrm{Exp}\, tA$ is a one-parameter subgroup with $\frac{d}{dt}(\mathrm{Exp}\, tA)|_{t=0} = A$. Hence $M_{n,n}(\mathbb{R})$ (resp. $M_{n,n}(\mathbb{C})$) is exactly the Lie algebra of $\mathrm{GL}(n, \mathbb{R})$ (resp. $\mathrm{GL}(n, \mathbb{C})$) and the above map, explicitly defined in terms of converging power series, is its exponential map. (This is the origin of the name "exponential map".) It is then not difficult to verify that

$$[A, B] = AB - BA,$$

for $A, B \in M_{n,n}(\mathbb{R})$ (resp. $M_{n,n}(\mathbb{C})$).

2. Suppose $A \in M_{n,n}(\mathbb{R})$ (resp. $M_{n,n}(\mathbb{C})$) and $\mathrm{Exp}\, tA \in O(n)$ (resp. $U(n)$) for all $t \in \mathbb{R}$, that is $\langle \mathrm{Exp}\, tA \cdot x, \mathrm{Exp}\, tA \cdot y \rangle = \langle x, y \rangle$ for all $t \in \mathbb{R}$. Then

$$\frac{d}{dt}\bigg|_{t=0} \langle \mathrm{Exp}\, tA \cdot x, \mathrm{Exp}\, tA \cdot y \rangle = \langle A \cdot x, y \rangle + \langle x, A \cdot y \rangle = 0.$$

Hence, A is skew symmetric (resp. skew hermitian). Actually $\mathrm{Exp}\, tA \subset O(n)$ (resp. $U(n)$) is equivalent to A being skew symmetric (resp. hermitian). Therefore the Lie subalgebra corresponding to the Lie subgroup $O(n) \subset \mathrm{GL}(n, \mathbb{R})$ (resp. $U(n) \subset \mathrm{GL}(n, \mathbb{C})$) is the Lie subalgebra of skew symmetric (resp. skew hermitian) matrices.

2. Lie Subgroups and the Fundamental Theorem of Lie

The study of one-parameter subgroups of a Lie group G enables us to obtain a linear object, namely, its Lie algebra $\mathfrak{G}$. It is undoubtedly a structure of much simpler type than that of the Lie group structure. However, the true value of such an "invariant" shall depend more on how powerful it is rather than how elementary its structure is. Therefore, our next topic of discussion is to apply this "newly gained" invariant to some basic problems of Lie groups in order to test its powerfulness. Our experiences both in abstract group theory and in Galois theory clearly indicate the importance of studying the subgroups of a given group. Therefore, it is natural to test its power on the problem of Lie subgroups.

Definition (H, ι) is called a Lie subgroup of a Lie group G if H is a Lie group and $\iota : H \to G$ is an injective differentiable homomorphism.

Caution!   The image set $\iota(H) \subset G$ may not be closed. For example $(\mathbb{R}, \iota)$ with $\iota(t) = (t, \sqrt{2}t) \bmod \mathbf{Z}^2$ is a Lie subgroup in $T^2 = \mathbb{R}^2/\mathbb{Z}^2$. But $\iota(\mathbb{R})$ is a dense subset in T^2.

Formulation of a Basic Testing Problem   Suppose $\iota : H \to G$ is a given Lie subgroup. Then the left cosets of H in G form a left-invariant foliation of G whose tangent subspace at $x \in G$ is exactly $dl_x(d\iota(\mathfrak{H}))$. Locally, one may choose a suitable coordinate neighborhood U of e such that the restriction of the above foliation to U is simply the foliation of "coordinate slices", namely, its leaves are given by

$$x^i = \text{const.}, \quad \dim H < i \le \dim G.$$

From here, it is easy to verify that $d\iota(\mathfrak{H}) \subset \mathfrak{G}$ is a Lie subalgebra, i.e. a subspace of $\mathfrak{G}$ closed under bracket operation. Therefore, the following problem on the uniqueness and the existence of connected Lie subgroups with a given Lie subalgebra as its Lie algebra is naturally a fundamental testing problem.

Problem   Let $\mathfrak{G}$ be the Lie algebra of G and $\mathfrak{H}$ be a Lie subalgebra of $\mathfrak{G}$. Does there always exist a connected Lie subgroup H with $\mathfrak{H}$ as its Lie algebra? Is such a connected Lie subgroup necessarily unique?

The following fundamental theorem of Lie provides the affirmative answer to the above problem and thus convincingly demonstrates the power of Lie algebras as an invariant for studying Lie groups.

Theorem 2.   *Let $\mathfrak{G}$ be the Lie algebra of a Lie group G and let $\mathfrak{H}$ be a Lie subalgebra of $\mathfrak{G}$. Then there exists a unique connected Lie subgroup (H, ι) which makes the following diagram commutative:*

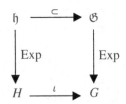

As expected, the proof of the above fundamental theorem of Lie relies heavily on the higher dimensional generalization of the existence and uniqueness theorems of ODE, namely, the Frobenius theorem on the complete integrability of involutive distributions. Therefore we shall first give a proof of the Frobenius Theorem and then deduce Theorem 2 from it. Let us first begin with a few needed definitions.

Definition   Let $X_1, \ldots, X_k$ be k smooth vector fields defined on an open neighborhood U such that $\{X_1(x), \ldots, X_k(x)\}$ is linearly independent for every $x \in U$. Set Δ_x equal to the span of $\{X_1(x), \ldots, X_k(x)\} \subset T_x M$. Then the k-plane field Δ on U which assigns the k-dimensional subspace Δ_x to each $x \in U$, is called a smooth k-dimensional distribution spanned by $\{X_1(x), \ldots, X_k(x)\}$.

Definition   A k-dimensional distribution Δ on M assigns a k-dimensional subspace Δ_x of $T_x M$ to each $x \in M$, namely, it is simply a k-plane field defined on M. It is called smooth if it can always be locally spanned be k smooth vector fields.

Definition   A k-dimensional smooth distribution Δ on M is called involutive if every set of local generating vector fields $\{X_1, \ldots, X_k\}$ always satisfies the following condition:

$$[X_i, X_j](x) \in \Delta_x \quad \forall\, x \in U \quad 1 \le i, j \le k,$$

or equivalently,

$$[X_i, X_j] = \sum_{l=1}^{k} f_{ij}^l X_l, \quad f_{ij}^l \in C^\infty(U).$$

It is easy to check that the above involutivity does not depend on the choice of generating vector fields, and hence it is a property of the distribution Δ.

Definition   A k-dimensional distribution Δ on M is called completely integrable if, to every point $x_0 \in M$, there always exists a suitable local coordinate neighborhood U such that $\Delta|U$ is spanned by $\frac{\partial}{\partial x^1}, \ldots, \frac{\partial}{\partial x^k}$.

Frobenius Theorem   *A k-dimensional distribution Δ on M is completely integrable if and only if it is involutive.*

Proof: The "only if" part is obvious. We shall only prove the "if" part by induction on the dimension k of the distribution. Notice that the starting point of $k = 1$ is essentially the usual existence-uniqueness theorem of ODE, namely, an everywhere non-vanishing smooth vector field X can always be locally expressed as $X = \frac{\partial}{\partial x^1}$ with respect to a suitably chosen local coordinate system. Therefore we begin our inductive proof by assuming that $k > 1$, $X_1 = \frac{\partial}{\partial x^1}$ and the above theorem already holds for smooth distribution of dimension $\leq k - 1$.

Let $\{\frac{\partial}{\partial x^1}, X_2, \ldots, X_k\}$ be a given set of local generating vector fields of an involutive distribution $\Delta|U$, and $x^1, \ldots, x^n$ are the local coordinate functions defined on U. Set

$$\tilde{X}_i = X_i - (X_i \cdot x^1) \cdot \frac{\partial}{\partial x^1}, \quad 2 \leq i \leq k.$$

Then $\{\frac{\partial}{\partial x^1}, \tilde{X}_2, \ldots, \tilde{X}_k\}$ is also a set of generating vector fields of $\Delta|U$ and $\tilde{X}_i x^1 \equiv 0$, $2 \leq i \leq k$. Let U_0 be the $(n-1)$-dimensional submanifold of U defined by $x^1 = 0$. Then the restrictions of $\{\tilde{X}_2, \ldots, \tilde{X}_k\}$ onto U_0 span an involutive distribution of dimension $k - 1$. Therefore, by the induction assumption, there exists a local coordinate system, say $(y^2, \ldots, y^n)$, such that for all $y \in U_0$,

$$\langle \tilde{X}_2(y), \ldots, \tilde{X}_k(y) \rangle = \left\langle \frac{\partial}{\partial y^2}, \ldots, \frac{\partial}{\partial y^k} \right\rangle.$$

To each point $p \in U$, let q be the unique point of intersection of U_0 with the x^1-curve passing through p, and $(y^2, \ldots, y^n)$ be the coordinates of q in U_0. Then $(x^1, y^2, \ldots, y^n) \leftrightarrow p$ constitutes a new local coordinate system of U and we shall show that for $x \in U$

$$\Delta_x = \left\langle \frac{\partial}{\partial x^1}, \frac{\partial}{\partial y^2}, \ldots, \frac{\partial}{\partial y^k} \right\rangle.$$

Or equivalently, what we need to show is that

$$\tilde{X}_i y^l \equiv 0, \quad 2 \leq i \leq k, \quad k + 1 \leq l \leq n.$$

Of course, one needs only to show that the restrictions of the above functions to every x^1-curve are all identically zero. Let Γ be an arbitrary x^1-curve and set

$$f_i^l(x^1) = \tilde{X}_i y^l|_\Gamma.$$

Then it follows from the involutivity of Δ that there exist smooth functions $\{h_{ij} : 2 \leq i, j \leq k\}$ such that

$$\left[\frac{\partial}{\partial x^1}, \tilde{X}_i\right] = \sum_{j=2}^{k} h_{ij} \tilde{X}_j, \quad 2 \leq i \leq k.$$

It is crucial to note that $[\frac{\partial}{\partial x^1}, \tilde{X}_i]x^1 = \frac{\partial}{\partial x^1}\tilde{X}_i x^1 - \tilde{X}_i 1 = 0 - 0 = 0$. Applying the above equations to y^l and then restricting to the x^1-curve Γ, one gets

$$\frac{d}{dx^1}f_i^l(x^1) = \left[\frac{\partial}{\partial x^1}, \tilde{X}_i\right]y^l\bigg|_{\Gamma} = \sum_{j=2}^{k} h_{ij}|_{\Gamma} \cdot f_j^l(x^1).$$

For fixed l, the above equations constitute a system of homogeneous first-order ODE and $\{f_i^l(x^1) : 2 \le i \le k\}$ is its unique set of solutions with zero initial values, i.e. $f_i^l(0) = 0$. It follows easily from the homogeneity that this unique set of solutions must be $\{f_i^l(x^1) \equiv 0\}$! This completes the proof of the Frobenius Theorem by induction. $\qquad\square$

Definition A k-dimensional submanifold $Y \subset M$ is said to be an integral submanifold of a k-dimensional smooth distribution Δ on M if $T_y Y = \Delta_y$ for all $y \in Y$.

Definition A connected integral submanifold Y is said to be a maximal integral submanifold of Δ if it cannot be properly contained in another connected integral submanifold of Δ.

Corollary *If Δ is an involutive distribution on M, then to each given point $x \in M$, there exists a unique maximal integral submanifold of Δ passing through x.*

Proof: Locally, the above theorem proves that integral submanifolds of an involutive Δ are simply the coordinate slices. Therefore one has the strongest possible local existence-uniqueness that two connected integral submanifolds with a single point in common can be pieced together to become a bigger one. Hence the unique maximal integral submanifold of Δ passing through a given point $x \in M$ is exactly the one obtained by pushing the above piecing together analytic continuation of the local coordinate slice of x to its utmost limit. $\qquad\square$

Proof of Theorem 2: Let $\mathfrak{H} \subset \mathfrak{G}$ be a given Lie subalgebra and $\{X_1, \ldots, X_k\}$ be an arbitrary basis of $\mathfrak{H}$. Interpret them as left-invariant vector fields. Then they, in fact, globally generate a left-invariant distribution $\Delta(\mathfrak{H})$ of dimension k on G. Since $\mathfrak{H}$ is assumed to be closed under bracket operation, $\Delta(\mathfrak{H})$ is involutive. Hence, it is completely integrable. Let H be the unique maximal integral submanifold of $\Delta(\mathfrak{H})$ passing through the identity. It follows from the left-invariance of $\Delta(\mathfrak{H})$ that $l_a(H) = a \cdot H$ is exactly the unique maximal integral submanifold of $\Delta(\mathfrak{H})$ passing through a. Let $h \in H$ be an arbitrary element of H. Then

$$e \in H \cap h^{-1} \cdot H \Rightarrow H = h^{-1} \cdot H \Rightarrow H = \bigcup_{h \in H} h^{-1}H = H^{-1}H.$$

Hence H is a connected Lie subgroup of G whose tangent space at e is exactly $\mathfrak{H}$. This proves the unique existences of a connected Lie subgroup of G corresponding to the given Lie subalgebra $\mathfrak{H}$ of $\mathfrak{G}$. □

Exercises   1. Show that a connected topological group G can always be generated by an arbitrary neighborhood of the identity. (To a given neighborhood U of e, choose another smaller one V with $V = V^{-1}$. Then $\bigcup_{n=1}^{\infty} V^n$ is an open subgroup of G.)

2. Show that there exists a sufficiently small neighborhood W of the origin in $\mathfrak{G}$ such that $\mathrm{Exp}\, W$ is a neighborhood of the identity in G, and moreover, the only subgroups of G contained in $\mathrm{Exp}\, W$ is the trivial one, $H = \{e\}$.

3. Lie Homomorphisms and Simply Connected Lie Groups

Next let us extend our general investigation of the relationship between Lie algebras and Lie groups to the case of Lie homomorphisms. Let G_1, G_2 be two connected Lie groups with $\mathfrak{G}_1$, $\mathfrak{G}_2$ as their Lie algebras. Suppose $\phi : \mathfrak{G}_1 \to \mathfrak{G}_2$ is a given Lie algebra homomorphism. Does there always exist a Lie group homomorphism Φ, such that $\phi = d\Phi|_e$, namely

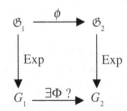

Does there exist Φ which makes the above diagram commutative?

As it turns out, the answer for the above problem is *not* universally affirmative. For example, if we take the simple case of $G_1 = S^1$ and $G_2 = \mathbb{R}$, both of their Lie algebras are $\mathbb{R}$ and

$$\mu_c : \mathbb{R} \to \mathbb{R} : t \mapsto ct, \quad c \in \mathbb{R}$$

are all the Lie algebra homomorphisms. However, the only Lie group homomorphism $\Phi : S^1 \to \mathbb{R}$ is the trivial one. ($\mathbb{R}$ contains no compact subgroups except the trivial one.) But if we interchange the positions, namely, $G_1 = \mathbb{R}, G_2 = S^1$, then it is not difficult to see that corresponding to each μ_c there is

$$\Phi_c : \mathbb{R} \to S^1 : t \mapsto e^{ict}$$

with $d\Phi_c|_e = \mu_c$.

What is the crucial point that makes the difference? Let us make use of Theorem 2 to help us to analyze the situation. Observe that the Lie algebra of $G_1 \times G_2$ is just $\mathfrak{G}_1 \oplus \mathfrak{G}_2$, and moreover, corresponding to each given Lie algebra homomorphism $\phi : \mathfrak{G}_1 \to \mathfrak{G}_2$, its graph

$$\Gamma(\phi) = \{(X, \phi(X)) : X \in \mathfrak{G}_1\}$$

is a Lie subalgebra of $\mathfrak{G}_1 \oplus \mathfrak{G}_2$. Hence, by Theorem 2, there exists a unique connected Lie subgroup $H \subset G_1 \times G_2$ with $\Gamma(\phi)$ as its Lie algebra. We need the following commutative diagram to put the whole situation in clear view.

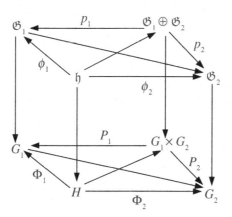

where the four vertical maps are Exp and

$$\phi_1 : \Gamma(\phi) \overset{\subseteq}{\to} \mathfrak{G}_1 \oplus \mathfrak{G}_2 \overset{p_1}{\to} \mathfrak{G}_1$$

is, by definition, invertible and $\phi = \phi_2 \circ \phi_1^{-1}$. Now here is the crucial point. Suppose that Φ_1 also happens to be invertible. Then $\Phi = \Phi_2 \circ \Phi_1^{-1}$ will clearly be the desired Lie group homomorphism. The fact that ϕ_1 is an iso-morphism, however, only implies that Φ_1 is a covering homomorphism. One needs the topological condition that G_1 is simply connected, i.e. $\pi_1(G_1) = 0$, to ensure that Φ_1 is also an isomorphism. Therefore one has the general existence of the corresponding Lie homomorphism for the special case that G_1 is a simply connected Lie group.

Theorem 3. *If G_1 is a simply connected Lie group, then to any given Lie algebra homomorphism $\phi : \mathfrak{G}_1 \to \mathfrak{G}_2$, there exists a unique Lie group homomorphism $\Phi : G_1 \to G_2$ such that $d\Phi_e = \phi$, namely*

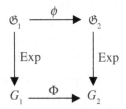

A theorem of Ado asserts that any finite dimensional (abstract) Lie algebra over $\mathbb{R}$ can be realized as (or rather, is isomorphic to) a Lie subalgebra of the Lie algebra of $\mathrm{GL}(n, \mathbb{R})$ for sufficiently large n. Therefore it follows from Theorem 2 that any finite dimensional (abstract) Lie algebra over $\mathbb{R}$ can be realized as the Lie algebra of a Lie group.

It is a well-known fact that every connected smooth manifold M has a unique universal covering manifold $\tilde{M}$, $f : \tilde{M} \to M$ is a covering map and $\pi_1(\tilde{M}) = 0$. A generic way of constructing $\tilde{M}$ directly from M is as follows:

Choose a fixed base point $x_0 \in M$. Let $P(M, x_0)$ be the set of all paths in M with x_0 as their initial point. Introduce the equivalence relation that $\gamma_1 \sim \gamma_2$ if they also have the same terminating point and they are homotopic with the end points stationary. Then $\tilde{M}$ is naturally bijective to $P(M, x_0)/\sim$.

In view of Theorem 3, it is quite natural to ask whether every finite dimensional Lie algebra over $\mathbb{R}$ can be realized as the Lie algebra of a simply connected Lie group. The following lemma provides the missing link for a proof of the affirmative answer to the above question.

Lemma   *Let G be a given connected Lie group and $h : \tilde{G} \to G$ is the universal covering manifold of G. Then there is a unique group structure on $\tilde{G}$ which makes $\tilde{G}$ into a Lie group and h into a Lie homomorphism.*

Proof: Consider $\tilde{G}$ as the space of equivalence classes of $P(G, e)$. One may define a natural, induced multiplication among elements of $P(G, e)$, namely, for the two paths

$$\gamma_i : [0, 1] \to G, \quad i = 1, 2,$$

one defines the product $\gamma_1 \cdot \gamma_2$ by the following formula:

$$(\gamma_1 \cdot \gamma_2)(t) = \gamma_1(t) \cdot \gamma_2(t).$$

It is easy to check that $\gamma_1 \sim \gamma_1'$ and $\gamma_2 \sim \gamma_2'$ implies $\gamma_1 \cdot \gamma_2 \sim \gamma_1' \cdot \gamma_2'$. Hence, the above multiplication induces a multiplication on $\tilde{G}$. From here, it is straightforward to check that the above multiplication makes $\tilde{G}$ into a Lie group and h into a Lie homomorphism, namely, it makes $h : \tilde{G} \to G$ into a covering Lie group. $\qquad\square$

Summarizing the discussion of this section, one may restate the results in terms of categorical language as follows. Let

1. LG be the category of Lie groups and Lie homomorphism,
2. LG_0 be the category of simply connected Lie groups and Lie homomorphisms,
3. LA be the category of Lie algebras and Lie algebra homomorphisms.

Then, the above results show that the linearization functor

$$\mathcal{L} : LG \to LA$$

becomes an isomorphism if restricted to the subcategory of LG_0, namely

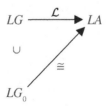

Exercises   1. Show that $\pi_1(G)$, G a Lie group, is necessarily commutative.

2. Show that a discrete normal subgroup of a connected Lie group G must be contained in the center of G.

3. Classify all connected Lie groups whose Lie algebras have trivial bracket operations, i.e. $[X, Y] \equiv 0$ for all $X, Y \in \mathfrak{G}$.

4. Adjoint Actions and Adjoint Representations

In the study of the structure of a given Lie group G, the major task lies in analyzing its "non-commutativity". It is intuitively advantageous to organize the non-commutativity of a Lie group G into the geometric object of its *adjoint action*, namely

$$\mathrm{Ad} : G \times G \to G, \quad (g, x) \mapsto gxg^{-1}.$$

As one shall see in later discussions, the study of the orbit structure of the adjoint transformation of G on itself is exactly the focal point of the whole structure theory of Lie groups.

Formally, the above action map $\mathrm{Ad}(g, x) = gxg^{-1}$ is a map of two "variables", namely, g and x. Therefore, in the spirit of the Lie algebra, one should look into its two stages of linearization as follows.

The First Stage   For each $g \in G$, $\mathrm{Ad}(g, \cdot) : G \to G$ is a Lie automorphism. Hence, there corresponds a Lie algebra automorphism, $\mathrm{Ad}_g : \mathfrak{G} \to \mathfrak{G}$ which makes the following diagram commutative

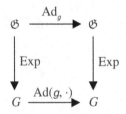

namely, $\mathrm{Exp}\, t\, \mathrm{Ad}_g(X) = g(\mathrm{Exp}\, tX)g^{-1}$. Therefore, one has a linear transformation group, $\mathrm{Ad}: G \times \mathfrak{G} \to \mathfrak{G}$, which maps G into the automorphism group of $\mathfrak{G}$, i.e.

$$\mathrm{Ad} : G \to \mathrm{Aut}(\mathfrak{G}) \subset \mathrm{GL}(\mathfrak{G}).$$

The Second Stage Let $\mathfrak{GL}(\mathfrak{G})$ be the Lie algebra of $GL(\mathfrak{G})$. Then the above Lie homomorphism, again, induces a Lie algebra homomorphism ad: $\mathfrak{G} \to \mathfrak{GL}(\mathfrak{G})$, which makes the following diagram commutative;

$$
\begin{array}{ccc}
\mathfrak{G} & \xrightarrow{\ \text{ad}\ } & \mathfrak{GL}(\mathfrak{G}) \\
\Big\downarrow{\scriptstyle \text{Exp}} & & \Big\downarrow{\scriptstyle \text{Exp}} \\
G & \xrightarrow{\ \text{Ad}\ } & GL(\mathfrak{G})
\end{array}
$$

Theorem 4. $\operatorname{ad}(X) \cdot Y = [X, Y]$, *for all* $X, Y \in \mathfrak{G}$.

Proof: By the above definitions, one has

$$\operatorname{Exp} sX \cdot \operatorname{Exp} tY \cdot \operatorname{Exp}(-sX)$$

$$= \operatorname{Exp} t[\operatorname{Ad}(\operatorname{Exp} sX) \cdot Y]$$

$$= \operatorname{Exp} t[\operatorname{Exp} s \cdot \operatorname{ad}(X) \cdot Y]$$

$$\equiv \operatorname{Exp} t[Y + s \cdot \operatorname{ad}(X) \cdot Y] \quad (\text{mod terms of order } \geq 3)$$

$$\equiv \operatorname{Exp} tY \cdot \operatorname{Exp} st \cdot \operatorname{ad}(X) \cdot Y \quad (\text{mod terms of order } \geq 3).$$

Hence

$$\operatorname{Exp} st \cdot [X, Y] \equiv \operatorname{Exp} sX \cdot \operatorname{Exp} tY \cdot \operatorname{Exp}(-sX) \cdot \operatorname{Exp}(-tY)$$

$$\equiv \operatorname{Exp} st \cdot \operatorname{ad}(X) \cdot Y \quad (\text{mod terms of order } \geq 3).$$

Therefore,

$$\operatorname{ad}(X) \cdot Y = [X, Y]. \qquad \square$$

Examples 1. $G = GL(n, \mathbb{R})$, $\mathfrak{G} = M_{n,n}(\mathbb{R})$.

In the special case, for each $g \in G$ and $X \in \mathfrak{G}$, one has

$$\operatorname{Exp} t \operatorname{Ad}(g)X = g \operatorname{Exp} tX g^{-1}$$

$$= g\left\{ I + tX + \frac{1}{2}(tX)^2 + \cdots + \frac{1}{k!} tX^k + \cdots \right\} g^{-1}$$

$$= I + tgXg^{-1} + \frac{t^2}{2}(gXg^{-1})^2 + \cdots + \frac{T^k}{k!}(gXg^{-1})^k + \cdots$$

$$= \operatorname{Exp} t(gXg^{-1}).$$

Therefore $\mathrm{Ad}(g) \cdot X = gXg^{-1}$. If we denote the birth certificate representation of $\mathrm{GL}(n, \mathbb{R})$ on $M_{n,1}(\mathbb{R}) \simeq \mathbb{R}^n$ by $\tilde{\rho}_n$, then the above adjoint representation is equivalent to $\tilde{\rho}_n \otimes_{\mathbb{R}} \tilde{\rho}_n^*$.

2. $G = \mathrm{GL}(n, \mathbb{C})$, $\mathfrak{G} = M_{n,n}(\mathbb{C})$.

Exactly the same reasoning will show that

$$\mathrm{Ad}(g)X = gXg^{-1},$$

for $g \in \mathrm{GL}(n, \mathbb{C})$ and $X \in M_{n,n}(\mathbb{C})$. And moreover, if we denote the birth certificate representation of $\mathrm{GL}(n, \mathbb{C})$ on $M_{n,1}(\mathbb{C})$ by $\tilde{\mu}_n$, then $\mathrm{Ad} = \tilde{\mu}_n \otimes_{\mathbb{C}} \tilde{\mu}_n^*$.

3. $G = O(n)$, $\mathfrak{G} = $ the space of skew-symmetric $n \times n$ matrices.

Again, one has for $g \in O(n)$ and $X \in \mathfrak{G}$

$$\mathrm{Ad}(g)X = gXg^{-1}.$$

If one denotes the birth certificate representation of $O(n)$ on $\mathbb{R}^n$ by ρ_n, i.e. $\rho_n = \tilde{\rho}_n|O(n)$, then

$$\mathrm{Ad}_{O(n)} = \Lambda^2 \rho_n$$

the restriction of the $O(n)$-conjugation to skew-symmetric matrices.

4. $G = U(n)$, $\mathfrak{G} = $ the space of skew-hermitian $n \times n$ matrices.

Again, one has

$$\mathrm{Ad}(g)X = gXg^{-1}.$$

Observe that every element $A \in M_{n,n}(\mathbb{C})$ can be uniquely expressed as the sum of its hermitian part and its skew hermitian part, namely, $A = \frac{1}{2}(A + A^*) + \frac{1}{2}(A - A^*)$; and moreover, B is hermitian if and only if iB is skew hermitian. Hence $\mathfrak{G} \otimes \mathbb{C} \simeq M_{n,n}(\mathbb{C})$.

Let $\mu_n = \tilde{\mu}_n|U(n)$. Then for $X \in \mathfrak{G}$

$$\mathrm{Ad}(g)X = gXg^{-1} \Rightarrow \mathrm{Ad}(g)(X + iY) = g(X + iY)g^{-1}.$$

Hence the complexification of the adjoint representation of $U(n)$ is exactly the above conjugation transformation of $U(n)$ on $M_{n,n}(\mathbb{C})$, namely,

$$\mathrm{Ad}_{U(n)} \otimes \mathbb{C} = \mu_n \otimes_{\mathbb{C}} \mu_n^*.$$

Lecture 3

Orbital Geometry of the Adjoint Action

Roughly speaking, the linear representation theory that we discussed in the first lecture is a kind of extrinsic linearization; and the basic Lie group theory that we discussed in the second lecture is a kind of intrinsic linearization. The former is based on the compactness of the group and the technique of averaging provided by the Haar integral; and the latter is based upon the differentiability of the group structure and technically relies on the existence-uniqueness theory of ordinary differential equations, in the form of Frobenius theorem on complete integrability of distributions. Therefore, in the case of compact connected Lie groups, one naturally expects that they can be combined to provide a rather satisfactory understanding of both the structural theory and the representation theory of this important family of groups. This shall be exactly the topic of our discussion for the next few lectures.

As it had already been pointed out in the previous lecture, the central issue in the study of group structure is the *non-commutativity* and it is advantageous to organize it in the form of *adjoint action*. Therefore, we shall begin our study of compact connected Lie groups by focusing our attention on the orbital geometry of the adjoint action, namely, the geometry of conjugacy classes. For example, $S^3 \cong \mathrm{SU}(2)$ is one of the simplest, non-commutative, compact connected Lie group, the orbital geometry of the conjugacy classes of S^3 had already been worked out in Lecture 1, which is exactly the crucial geometric input that enable us to apply the character theory of Frobenius–Schur to obtain a neat classification of the complex irreducible representations of S^3. In fact, this simple special case will serve as a good prototype for the general theory of compact connected Lie groups.

1. Bi-Invariant Riemannian Structure on a Compact Connected Lie Group and the Maximal Tori Theorem of É. Cartan

In this lecture, we shall always assume that G is a compact connected Lie group and $\mathfrak{G}$ is its Lie algebra. The compactness of G ensures the existence of adjoint-invariant inner products on $\mathfrak{G}$. Fix such an invariant inner product on $\mathfrak{G}$ and then choose an orthonormal basis of $\mathfrak{G}$, say $\{X_i; 1 \leq i \leq \dim \mathfrak{G}\}$. Let $\tilde{X}_i$ be the left invariant vector field on G with X_i as its value at the identity e. This frame field $\{\tilde{X}_i; 1 \leq i \leq \dim \mathfrak{G}\}$ uniquely determines a Riemannian structure on G such that $\{\tilde{X}_i(x); 1 \leq i \leq \dim \mathfrak{G}\}$ is an orthonormal basis of $T_x G$ for all x in G.

Lemma 1.　*The above Riemannian metric on G is bi-invariant, namely, both left and right translations are isometries.*

Proof:　Observe that the inner product of a vector space is uniquely determined by one of its orthonormal basis. Therefore, a linear map $A : V \to W$ is an isometry if and only if A maps an orthonormal basis of V to an orthonormal basis of W. Let l_a (resp. r_a) be the left (resp. right) translation of G by a, i.e. $l_a(x) = a \cdot x$ (resp. $r_a(x) = x \cdot a$). Then, the left invariance of $\tilde{X}_i$ simply means that $dl_a : T_x G \to T_{ax} G$ maps $\tilde{X}_i(x)$ to $\tilde{X}_i(ax)$. Hence, all left translations, l_a, $a \in G$, are obviously isometric. Next, let us consider the following diagram of linear maps.

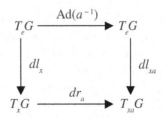

It is commutative because

$$xa(a^{-1}ga) = xga \Rightarrow l_{xa} \circ \mathrm{Ad}(a^{-1}) = r_a \circ l_x.$$

Therefore, the fact that dl_x, dl_{xa} and $\mathrm{Ad}(a^{-1})$ are all isometries implies that dr_a is also an isometry. $\qquad\square$

From now on, a compact connected Lie group G is always assumed to be equipped with such a bi-invariant Riemannian metric. Hence, in particular, the adjoint action of G on the Riemannian manifold G is an *isometric* transformation group whose orbits are exactly the conjugacy classes of the group G. The key result in the geometric structure of conjugacy classes of a compact connected Lie group G is the maximal tori theorem of É. Cartan. Recall that the group of unit complexes, i.e., the circle group S^1, is the only one-dimensional compact connected Lie group, and moreover, the products of several copies of the circle group S^1 are the only *commutative*, compact connected Lie groups. The product of k copies of S^1 is called a torus group of rank k and shall be denoted by T^k or simply by T if its rank does not need to be specified.

Definition   A torus subgroup $T \subset G$ is called a maximal torus of G if it cannot be properly contained in any other torus subgroup of G.

Lemma 2.   *Let T be a torus subgroup of G and $F(T,\mathfrak{G})$ (resp. $F(T,G)$) be the fixed point set of adjoint action of T on $\mathfrak{G}$ (resp. G). Then T is a maximal torus of G if and only if either $\dim F(T,\mathfrak{G}) = \dim T$ or $F(T,G)$ contains T as one of its connected components.*

Proof: Let $\mathfrak{M}$ be the Lie algebra of T. Then it follows from the definition of the adjoint action of T on $\mathfrak{G}$ that

$$F(T,\mathfrak{G}) \supset \mathfrak{M}.$$

If T is not maximal, say T is properly contained in another torus subgroup T_1 with $\mathfrak{M}_1$ as its Lie algebra, then

$$F(T,\mathfrak{G}) \supset \mathfrak{M}_1 \Rightarrow \dim F(T,\mathfrak{G}) \geq \dim \mathfrak{M}_1 > \dim T.$$

Conversely, suppose $\dim F(T,\mathfrak{G}) > \dim T$. Then, there exists

$$X \in F(T,\mathfrak{G})\backslash\mathfrak{M}$$

and hence $\mathfrak{M}_1 = \langle X, \mathfrak{M} \rangle$ is a Lie subalgebra of $\mathfrak{G}$ with identically zero bracket operation. By Theorems 2.2 and 2.3, there exists a unique commutative connected Lie group H with $\mathfrak{M}_1$ as its Lie algebra. The closure of H is certainly a torus subgroup of G, i.e. a commutative compact connected subgroup of G, which properly contains T. Hence T is not a maximal torus. This completes the proof that

$$T \text{ is a maximal torus } \Leftrightarrow \dim F(T, \mathfrak{G}) = \dim T.$$

The second condition is closely related to the above one because the identity component of $F(T, G)$ is clearly a Lie subgroup of G whose Lie algebra is exactly $F(T, \mathfrak{G})$. $\qquad\square$

Examples   1. The subgroup of unit complexes, S^1, is a maximal torus in the group of unit quaternions S^3.

2.

$$SO(2) = \left\{ \begin{pmatrix} \cos\theta & -\sin\theta & 0 \\ \sin\theta & \cos\theta & 0 \\ 0 & 0 & 1 \end{pmatrix} ; 0 \le \theta < 2\pi \right\}$$

is a maximal torus of $SO(3)$.

3.

$$T^n = \left\{ \begin{pmatrix} e^{i\theta_1} & & & & & \\ & e^{i\theta_2} & & & & \\ & & \ddots & & & \\ & & & e^{i\theta_j} & & \\ & & & & \ddots & \\ & & & & & e^{i\theta_n} \end{pmatrix} ; 0 \le \theta_j < 2\pi \right\}.$$

is a maximal torus of $U(n)$.

4. Let $T \subset U(n)$ be an arbitrary torus subgroup of $U(n)$. Then it follows from a corollary of the Schur Lemma that the above unitary representation, $(T, \mathbb{C}^n)$, splits into the direct sum of one-dimensional ones. Therefore, there exists a suitable orthonormal basis $\{\mathbf{b}_j; 1 \le j \le n\}$ which consists of common eigenvectors of all elements of T. Let $\{\mathbf{e}_j; 1 \le j \le n\}$ be the canonical

basis of $\mathbb{C}^n$, i.e., $\mathbf{e}_j = (0, \ldots, 0, 1, 0, \ldots, 0)$ and set B be the element of $U(n)$ with $B(\mathbf{e}_j) = \mathbf{b}_j$, $1 \leq j \leq n$. Then, for each $A \in T$, $1 \leq j \leq n$,

$$B^{-1}AB(\mathbf{e}_j) = B^{-1}A(\mathbf{b}_j) = B^{-1}(\lambda_j \mathbf{b}_j) = \lambda_j \mathbf{e}_j, \qquad |\lambda_j| = 1,$$

namely, one has

$$B^{-1}TB \subset T^n.$$

Exercises   1. Show that the subgroups in the above Examples 1–3 are indeed maximal torus in the respective groups.

2. Let $A \in U(n)$ be an arbitrary element in $U(n)$. Show that there always exists a suitable $B \in U(n)$ such that $B^{-1}AB \in T^n$, i.e. $B^{-1}AB$ is a diagonal unitary matrix.

3. Exhibits a maximal torus of SO(4).

Theorem 1 (É. Cartan). *Let T be a maximal torus of G. Then T intersects every conjugacy class of G, i.e. every element $g \in G$ is conjugate to a suitable element in T.*

Proof: Let $\varphi = \mathrm{Ad}\,|T$ be the restriction of the adjoint representation of G to T and $\mathfrak{h} \subset \mathfrak{G}$ be the Lie algebra of T. Since T is a maximal torus, it follows from Lemma 2 that $F(T, \mathfrak{G}) = \mathfrak{h}$. Recall that every *complex* irreducible representation of a torus group must be one-dimensional, it follows readily that every non-trivial *real* irreducible representation of a torus group is always two-dimensional. Hence

$$\varphi = \dim \mathfrak{h} \cdot 1 \oplus \varphi_1 \oplus \cdots \oplus \varphi_l,$$

where 1 denotes the one-dimensional trivial representation and φ_j, $1 \leq j \leq l$, are non-trivial homomorphisms of T onto SO(2). Therefore, $\ker(\varphi_j)$, $1 \leq j \leq l$, are all *codimension one* closed Lie subgroup of T; the complement of their union, $\bigcup \ker(\varphi_j)$, is an open dense submanifold of T, say denoted by W.

Let $t_0 \in W$ be an arbitrary element in W. Then each $\varphi_j(t_0)$ is a non-trivial rotation and hence $F(\varphi(t_0), \mathfrak{G}) = \mathfrak{h}$. Let $G_{t_0} = \{g \in G; g t_0 g^{-1} = t_0\}$ be the centralizer of t_0 and $\mathrm{Exp}\,sX$ be an arbitrary one-parameter subgroup of G_{t_0}. Then

$$\text{Exp}\, sX = t_0 \,\text{Exp}\, sX t_0^{-1} = \text{Exp}\, s\,\text{Ad}(t_0)X, \qquad \forall\, s \in \mathbb{R}$$

$$\Rightarrow X \in F(\varphi(t_0), \mathfrak{G}) = \mathfrak{h}.$$

Hence, the connected component of the identity of G_{t_0} is equal to T and, of course, $\dim G(t_0) = \dim G - \dim G_{t_0} = \dim G - \dim T$, namely, the conjugacy class of t_0, i.e. $G(t_0)$, and the maximal torus T are submanifolds of complementary dimensions. The key geometric fact that the entire proof is based upon is that T and $G(t_0)$ intersect *perpendicularly* and *transversally*! We shall prove the above fact by analyzing the action of T on $T_{t_0}G$.

Observe that l_{t_0} commutes with the conjugation of t, $t \in T$, namely, $l_{t_0} \circ \sigma_t(x) = t_0 t x t^{-1} = t t_0 x t^{-1} = \sigma_t \circ l_{t_0}(x)$ for all $x \in G$. Therefore, dl_{t_0} is an *equivariant* linear map of $\mathfrak{G} = T_e G$ onto $T_{t_0}G$ with respect to the induced T-actions. Recall that

$$\mathfrak{G} = \mathfrak{h} \oplus \mathfrak{h}^{\perp}$$

with $\varphi|\mathfrak{h} = \dim \mathfrak{h} \cdot 1$ and $\varphi|\mathfrak{h}^{\perp} = \varphi_1 \oplus \cdots \oplus \varphi_l$. It is clear that $dl_{t_0}(\mathfrak{h})$ is exactly the tangent space of T at t_0. Hence, in order to show that the tangent space of $G(t_0)$ at t_0 is exactly the orthogonal complement to that of T, it suffices to prove that the induced T-action on that of $G(t_0)$ contains no fixed directions. This is an easy corollary of the following simple but useful lemma. $\square$

Lemma 3.   *Let H be a compact Lie subgroup of G. Then the induced H-action on the tangent space of G/H at the based point, $T_0(G/H)$, is equivalent to the restriction of the adjoint H-action on $\mathfrak{G}$ to the (orthogonal) complement of $\mathfrak{h}$.*

Proof: Let $\mathfrak{G}$ be the Lie algebra of G equipped with an Ad_H-invariant inner product and $\mathfrak{h}^{\perp}$ be the orthogonal complement of the Lie subalgebra $\mathfrak{h}$. Let $p : G \to G/H$ be the canonical projection, i.e., $p(x) = x \cdot H \in G/H$. Observe that,

$$p \circ \sigma_h(x) = p(hxh^{-1}) = hxh^{-1} \cdot H = hx \cdot H = l_h(x \cdot H),$$

for all $h \in H$, namely, p is, in fact, an H-equivariant differentiable map with respect to the adjoint action of H on G and the left translation of H on G/H. Therefore,

$$dp_e : \mathfrak{G} = \mathfrak{h} \oplus \mathfrak{h}^{\perp} \to T_0(G/H)$$

maps $\mathfrak{h}^{\perp}$ isomorphically onto $T_0(G/H)$ as H-linear spaces. $\square$

Summarizing the above discussions, we have already obtained the following key geometric facts, namely

(i) T is a connected component of $F(T, G)$ and hence it is a *totally geodesic submanifold* of G. [Recall that the fixed point set of an isometric transformation group is always a totally geodesic submanifold. It is a direct consequence of the uniqueness of geodesic with given initial point and direction.]

(ii) The tangent spaces of T and $G(t_0)$ at t_0 are orthogonal complements of each other.

Based upon the above two facts, it is then an easy matter to complete the proof of Theorem 1 as follows.

Let $G(y)$ be any other G-orbit, i.e. conjugacy class. Then $G(t_0)$ and $G(y)$ are two compact submanifolds in the complete Riemannian manifolds G. Hence, by Hopf–Rinow Theorem, there always exists a geodesic interval, say $\overline{x_1 y_1}$, which realizes the shortest distance between them, and therefore, it must be perpendicular to both. Let g be a suitable element of G such that $g(x_1) = g x_1 g^{-1} = t_0$. Then $g(\overline{x_1 y_1}) = \overline{t_0 g(y_1)}$ is again a geodesic interval which is also perpendicular to both. Therefore, by (i) and (ii), the whole geodesic interval $\overline{t_0 g(y_1)}$ lies in T and hence $g(y_1) = g y_1 g^{-1} \in T \cap G(y)$. This completes the proof of Theorem 1.

Remark   In fact, the above proof actually provides much more informations on the orbital geometry of the adjoint action other than just the intersection property stated in Theorem 1. For examples, the following useful geometric facts are already included in the above proof.

(1) For every element $t_0 \in W$, $G(t_0)$ and T are of complementary dimensions and they intersect perpendicularly and transversally at t_0.

(2) For each element $t_1 \in \bigcup \ker(\varphi_j)$, $\dim G_{t_1} = \dim F(\varphi(t_1), \mathfrak{G}) \geq \dim \mathfrak{h} + 2$. Hence

$$\dim G(t_1) \leq \dim G - \dim T - 2,$$

$$\dim \left\{ \bigcup \ker(\varphi_j) \right\} = \dim T - 1,$$

$$\dim \cup \left\{ G(t_1); t_1 \in \bigcup \ker(\varphi_j) \right\} \leq \dim G - 3.$$

(3) In fact, to an arbitrary, fixed top dimensional orbit such as $G(t_0)$, the totality of maximal tori of G can be characterized as the set of complete,

totally geodesic normal submanifolds. Therefore, any two maximal tori of G are mutually conjugate. [Suppose T_1, T_2 are respectively such normal submanifolds of $G(t_0)$ at x_1, x_2 and $x_2 = gx_1g^{-1}$. Then gT_1g^{-1} and T_2 are both such normal submanifolds of $G(t_0)$ at x_2 and hence $gT_1g^{-1} = T_2$.]

Corollary 1. *All maximal tori of a compact connected Lie group G are mutually conjugate. [The common rank of maximal tori of G is defined to be the rank of G.]*

Corollary 2. *Let S be a torus subgroup of G and $Z_G(S)$ be the centralizer of S in G. Then $Z_G(S)$ is equal to the union of all maximal tori of G containing S (and hence it is connected), namely,*

$$Z_G(S) = \bigcup \{T; T \supset S\}$$

$$= \bigcup \{T; T \supset S \text{ and maximal}\}.$$

Proof: Clearly, $\bigcup \{T; T \supset S\} \subset Z_G(S)$ and

$$\bigcup \{T; T \supset S\} = \bigcup \{T; T \supset S \text{ and maximal}\}.$$

Hence, one need only to show that every $x \in Z_G(S)$ is contained in a maximal torus $T \supset S$.

Let H be the subgroup generated by $\{x, S\}$ and $\bar{H}$ be its closure; $\bar{H}$ is clearly Abelian and compact. If it is also connected, then $\bar{H}$ is a torus and we have nothing to prove. Next let us consider the case that $\bar{H}$ is disconnected. Let $\bar{H}_0$ be its identity component and $\bar{H}/\bar{H}_0$ be the quotient group which is generated by $x \cdot \bar{H}_0$, namely, $\bar{H} \cong \mathbb{Z}_l \times \bar{H}_0$ where $\bar{H}_0$ is a torus group and $\mathbb{Z}_l$ is a cyclic group of order l. Choose a suitable element a in $\bar{H}_0$ such that the cyclic group generated by a is dense in $\bar{H}_0$ (such a "topological generator" always exists for torus group, a theorem of Kronecker). Let b be a generator of $\mathbb{Z}_l$ and $c \in \bar{H}_0$ with $c^l = a$. Then $(b \cdot c)^l = b^l \cdot c^l = a$ and hence the cyclic group generated by $b \cdot c$, $\langle b \cdot c \rangle$, is dense in the whole $\bar{H}$, namely,

$$\overline{\langle b \cdot c \rangle} \supset \overline{\langle a \rangle} = \bar{H}_0 \Rightarrow b \in \overline{\langle b \cdot c \rangle} \Rightarrow \overline{\langle b \cdot c \rangle} = \bar{H}.$$

Now, by Theorem 1, $b \cdot c$ is contained in a maximal torus T and hence $x \in \bar{H} = \overline{\langle b \cdot c \rangle} \subset T$. This proves that

$$\bigcup \{T; T \supset S \text{ and maximal}\} \supset Z_G(S),$$

and thus

$$Z_G(S) = \bigcup \{T; T \supset S \text{ and maximal}\},$$

which is clearly connected. □

Corollary 3. $Z_G(T) = T$ *for a maximal torus T.*

Proof: By Lemma 2, T is equal to the identity component of $F(T, G) = Z_G(T)$. Hence the connectedness of $Z_G(T)$ implies $Z_G(T) = T$. □

Definition $W(G) = N_G(T)/T$ is called the Weyl group of G, where $N_G(T)$ is the normalizer of T in G.

Remarks (i) The restriction of the adjoint action map to $N_G(T) \times T$ naturally induces an action map of $W(G) \times T \to T$, namely,

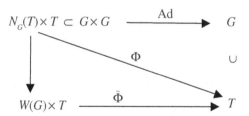

(ii) A torus group of rank k, $T \cong (\mathbb{R}/\mathbb{Z})^k \cong \mathbb{R}^k/\mathbb{Z}^k$. Therefore, the automorphism group of T is given by the group of invertible integral matrices of rank k, i.e. $\mathrm{Aut}(T) \cong GL(k, \mathbb{Z})$.

(iii) The action map $\Phi : N_G(T) \times T \to T$ induces a homomorphism $\varphi : N_G(T) \to \mathrm{Aut}(T)$ with $\ker(\varphi) = Z_G(T) = T$. Therefore, its effective quotient gives an injective map $\tilde{\varphi} : W(G) \to \mathrm{Aut}(T) \cong Gl(k, \mathbb{Z})$. Hence, $W(G)$ is a compact subgroup in a discrete group, namely, a finite group.

Weyl Reduction   We shall use G/Ad to denote the orbit space of the adjoint action on G, namely, the quotient space of conjugacy classes of G. The maximal tori theorem proves that a maximal torus T intersects every conjugacy class and hence the composition of $T \subset G \to G/\mathrm{Ad}$ is surjective. Moreover, it clearly factors through T/W, the orbit space of the Weyl group action on T, namely, one has the following commutative diagram.

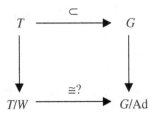

Naturally, one would like to know whether the above surjection is also injective?

Theorem 2.   *Both $T/W \to G/\mathrm{Ad}$ and $\mathfrak{h}/W \to \mathfrak{G}/\mathrm{Ad}$ in the following commutative diagrams are bijective, namely,*

$$
\begin{array}{ccc}
T & \xrightarrow{\subset} & G \\
\downarrow & & \downarrow \\
T/W & \xrightarrow{\cong} & G/\mathrm{Ad}
\end{array}
\qquad
\begin{array}{ccc}
\mathfrak{h} & \xrightarrow{\subset} & \mathfrak{G} \\
\downarrow & & \downarrow \\
\mathfrak{h}/W & \xrightarrow{\cong} & \mathfrak{G}/\mathrm{Ad}
\end{array}
$$

Proof: Since $F(T,G) = Z_G(T) = T$ [Corollary 3],

$$T \cap G(x_0) = F(T, G(x_0)), \qquad x_0 \in T$$

for any given conjugacy class $G(x_0)$. Let $x_1 = g x_0 g^{-1}$ be another point of $F(T, G(x_0))$. Then

$$T \subset G_{x_1} = g G_{x_0} g^{-1} \Rightarrow T \quad \text{and} \quad g^{-1} T g \subset G_{x_0},$$

namely, both T and $g^{-1}Tg$ are maximal tori of G_{x_0}. Hence, by Corollary 1, there exists $y \in G_{x_0}$ such that

$$yTy^{-1} = g^{-1}Tg \Rightarrow (gy)^{-1}Tgy = T.$$

Therefore, $gy \in N_G(T)$ and $x_1 = gx_0 g^{-1} = gyx_0 y^{-1} g^{-1}$ ($y \in G_{x_0}$ implies $yx_0 y^{-1} = x_0$), namely x_0 and x_1 are on the same W-orbit. This proves the injectivity and hence the bijectivity of $T/W \rightarrow G/\mathrm{Ad}$.

Since both $(W, \mathfrak{h})$ and $(G, \mathfrak{G})$ are respectively the local linearization of (W, T) and (G, G) at the identity, the injectivity of $\mathfrak{h}/W \rightarrow \mathfrak{G}/\mathrm{Ad}$ follows directly from that of the former. $\qquad\square$

2. Root System and Weight System

The combination of the above Theorem 1 and the character theory of Frobenius–Schur enable us to reduce the study of representations of a compact connected Lie group G to that of their restrictions to a maximal torus T.

Basic Fact 1. *Two representations of G, φ and ψ, are equivalent if and only if their restrictions to a maximal torus T, i.e. $\varphi|T$ and $\psi|T$, are equivalent, namely*

$$\varphi \sim \psi \Leftrightarrow \varphi|T \sim \psi|T.$$

Proof: Recall that the character functions of representations always take constant values on each conjugacy class and T intersects every conjugacy class. Therefore,

$$\chi_\varphi = \chi_\psi \Leftrightarrow \chi_\varphi|T = \chi_\psi|T,$$

and hence,

$$\varphi \sim \psi \quad \overset{\text{Thm 1.2}}{\Longleftrightarrow} \quad \chi_\varphi = \chi_\psi$$

$$\Updownarrow \text{Thm 3.1}$$

$$\varphi|T \sim \psi|T \quad \overset{\text{Thm 1.2}}{\Longleftrightarrow} \quad \chi_\psi|T = \chi_\psi|T. \qquad\square$$

Observe that the complex representations of a torus group T always splits into the direct sum of one-dimensional ones, the above reduction is, indeed, a rather advantageous one. Let φ be a given complex representation of G, T be a maximal torus of G and $\mathfrak{h}$ be the Lie algebra of T. Let

$$\varphi|T = \varphi_1 \oplus \varphi_2 \oplus \cdots \oplus \varphi_n, \quad n = \dim \varphi$$

be the splitting of $\varphi|T$ into one-dimensional representations, namely, each φ_i is a one-dimensional unitary representation. Recall that $\mathfrak{h}$ is simply a real vector space of dimension $k = \dim T = rk(G)$ and $\varphi_j : T \to U(1) \simeq S^1$ is uniquely determined by $\tilde{\varphi}_j = d\varphi_j|_e$, namely,

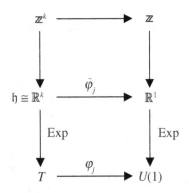

where $\tilde{\varphi}_j$ is an integral linear functional of $\mathfrak{h}$ ("integral" means $\tilde{\varphi}_j(\mathbb{Z}^k) \subset \mathbb{Z}$) which is an integral element of the dual space $\mathfrak{h}^*$ with respect to a specified basis.

Definition of Weight System   The weight system of a complex representation φ of G is defined to be the collection of the above integral linear functionals $\{\tilde{\varphi}_j; 1 \leq j \leq n\}$. It is a set of integral elements in $\mathfrak{h}^*$ with multiplicities.

The weight system of a real representation of G is defined to be the weight system of its complexification.

Remark   The weight system of φ is a complete set of invariants of φ and it is simply a convenient book-keeping device of φ. We shall use the notation $\Omega(\varphi)$ to denote the weight system of φ and $m(\omega, \varphi)$ or simply $m(\omega)$ to denote the multiplicity of ω in $\Omega(\varphi)$.

Definition of Root System   In the special case that φ is the adjoint representation of G, its system of non-zero weights is called the *root system of G*.

Remark   The multiplicity of zero weight in $\Omega(\mathrm{Ad} \otimes \mathbb{C})$ is equal to the rank of G, and the multiplicity of every non-zero weight in $\Omega(\mathrm{Ad} \otimes \mathbb{C})$ will be proved to be 1 in the next section. Therefore, it is convenient to *exclude* the zero weight in the definition of the root system of G, for it then becomes a set of uniform multiplicities equal to 1. Hence the root system of G is, in fact, just a set! We shall use the notation $\Delta(G)$ to denote the root system of G.

Basic Fact 2. *The weight system $\Omega(\varphi)$ and the character function $\chi_\varphi|T = \chi_{\varphi|T}$ are both complete invariants of $\varphi|T$, and hence also of φ itself. They are clearly related as follows.*

Let H be a generic element in $\mathfrak{h}$. Then it follows from the following diagram

$$
\begin{array}{ccc}
H \in \mathfrak{h} & \xrightarrow{\ \tilde{\varphi}_j\ } & \mathbb{R} \ \ni \ t \\
\Big\downarrow{\scriptstyle \mathrm{Exp}} & & \Big\downarrow{\scriptstyle \mathrm{Exp}} \quad \Big\downarrow \\
\mathrm{Exp}\,H \in T & \xrightarrow{\ \varphi_j\ } & U(1) \ \ni \ e^{2\pi i t}
\end{array}
$$

that $\chi_{\varphi_j}(\mathrm{Exp}\,H) = e^{2\pi i \tilde{\varphi}_j(H)}$. Hence

$$
\chi_\varphi(\mathrm{Exp}\,H) = \sum \chi_{\varphi_j}(\mathrm{Exp}\,H)
$$

$$
= \sum_{\omega \in \Omega(\varphi)} e^{2\pi i \omega(H)} \quad (\textit{sum with multi.}).
$$

Basic Fact 3. *One has the following convenient formulas for the character functions, namely,*

(i) $\chi_{\varphi \oplus \psi} = \chi_\varphi + \chi_\psi$,
(ii) $\chi_{\varphi \otimes \psi} = \chi_\varphi \cdot \chi_\psi$,
(iii) $\chi_{\varphi^*} = \bar{\chi}_\varphi$.

Correspondingly, one has the following useful relationships among the weight systems, namely,

(i) $\Omega(\varphi \oplus \psi) = \Omega(\varphi) \cup \Omega(\psi)$, *(with multi.)*
(ii)

$$
\Omega(\varphi \otimes \psi) = \Omega(\varphi) + \Omega(\psi)
$$

$$
= \{\omega_1 + \omega_2; \omega_1 \in \Omega(\varphi), \omega_2 \in \Omega(\psi)\}, \quad (\textit{with multi.})
$$

(iii) $\Omega(\varphi^*) = \widetilde{\Omega(\varphi)} = \{-\omega; \omega \in \Omega(\varphi)\}$ *(with multi.)*.

Examples   1. Let $G = S^3$. Then $S^1 = \{e^{2\pi i\theta}\}$ is a maximal torus of S^3. Let φ_k be the irreducible representation of dimension $k+1$. Then

$$\chi_{\varphi_k}(e^{2\pi i\theta}) = e^{2\pi ik\theta} + e^{2\pi i(k-2)\theta} + \cdots + e^{-2\pi ik\theta}.$$

Hence,

$$\Omega(\varphi_k) = \{k\theta, (k-2)\theta, \ldots, -k\theta\}.$$

Moreover, since $\mathrm{Ad} \otimes \mathbb{C} = \varphi_2$,

$$\Delta(S^3) = \{2\theta, -2\theta\}.$$

2. Let $G = U(n)$. Then

$$T = \left\{ \begin{pmatrix} e^{2\pi i\theta_1} & & & \\ & e^{2\pi i\theta_2} & & \\ & & \ddots & \\ & & & e^{2\pi i\theta_n} \end{pmatrix} \right\}$$

is a maximal torus of $U(n)$. Let μ_n be the birth certificate representation of the $U(n)$ action on $M_{n,1}(\mathbb{C})$. Then

$$\mu_n|T = \varphi_1 \oplus \varphi_2 \oplus \cdots \oplus \varphi_n,$$

where $\tilde{\varphi}_j = \theta_j$. Hence

$$\Omega(\mu_n) = \{\theta_j; 1 \le j \le n \quad \text{with } m(\theta_j) = 1\},$$

and correspondingly

$$\chi_{\mu_n}|T = \sum_{j=1}^{n} e^{2\pi i\theta_j}.$$

By Basic Fact 3,

$$\begin{cases} \Omega(\mu_n^*) = \{-\theta_j; 1 \le j \le n \quad \text{with } m(-\theta_j) = 1\}, \\ \chi_{\mu_n^*}|T = \displaystyle\sum_{j=1}^{n} e^{-2\pi i\theta_j}. \end{cases}$$

3. The complexification of $\mathrm{Ad}_{U(n)}$ is equal to $\mu_n \otimes \mu_n^*$. Therefore

$$\chi_{\mathrm{Ad}\otimes\mathbb{C}}|T = \left(\sum_{j=1}^{n} e^{2\pi i\theta_j}\right) \cdot \left(\sum_{k=1}^{n} e^{-2\pi i\theta_k}\right)$$

$$= n + \sum_{j \ne k} e^{2\pi i(\theta_j - \theta_k)},$$

and hence

$$\Delta(U(n)) = \{(\theta_j - \theta_k), 1 \le j \ne k \le n\}.$$

4. $G = SO(3)$. Then

$$SO(2) = \left\{ \begin{pmatrix} \cos 2\pi\alpha & -\sin 2\pi\alpha & 0 \\ \sin 2\pi\alpha & \cos 2\pi\alpha & 0 \\ 0 & 0 & 1 \end{pmatrix} \right\} \quad \text{is a maximal torus.}$$

The adjoint representation of S^3 is a covering homomorphism:

$$S^3 \overset{\mathrm{Ad}}{\to} SO(3), \ker = \{\pm 1\},$$

whose restriction to S^1 is a two-fold winding, namely,

$$S^1 \to SO(2), \quad e^{2\pi i\theta} \mapsto \begin{pmatrix} \cos 4\pi\theta & -\sin 4\pi\theta & 0 \\ \sin 4\pi\theta & \cos 4\pi\theta & 0 \\ 0 & 0 & 1 \end{pmatrix}.$$

Every complex irreducible representation of $SO(3)$ pulls back to that of S^3 with $\ker \supset \{\pm 1\}$; and conversely, every complex irreducible representation of S^3 with $\ker \supset \{\pm 1\}$ can also be pushed to that of $SO(3)$. It is easy to see that $\ker(\varphi_k) \supset \{\pm 1\}$ if and only if k is even. Therefore to each odd dimension $2l + 1$ there exists a unique complex irreducible representation of $SO(3)$, ψ_l such that

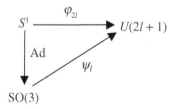

whose weight system is given as follows:

$$\Omega(\psi_l) = \{j \cdot \alpha, -l \le j \le l\}.$$

Notice that $\Delta(S^3) = \{\pm 2\theta\}$, $\Delta(SO(3)) = \{\pm\alpha\}$ and $2\theta \leftrightarrow \alpha$ in the covering map.

5. $G = SO(2l)$. Then

$$T = \left\{ \begin{pmatrix} \begin{array}{|cc|} \hline \cos 2\pi\theta_1 & -\sin 2\pi\theta_1 \\ \sin 2\pi\theta_1 & \cos 2\pi\theta_1 \\ \hline \end{array} & & \\ & \ddots & \\ & & \begin{array}{|cc|} \hline \cos 2\pi\theta_l & -\sin 2\pi\theta_l \\ \sin 2\pi\theta_l & \cos 2\pi\theta_l \\ \hline \end{array} \end{pmatrix} \right\}$$

$$\cong [SO(2)]^l$$

is a maximal torus of $SO(2l)$. Let ρ_{2l} be the birth certificate representation of $SO(2l)$ on $M_{2l,1}(\mathbb{R})$. Then

$$\rho_{2l}|T = \psi_1 \oplus \psi_2 \cdots \oplus \psi_l,$$

where $\psi_j : T \to SO(2)$ by the projection to its jth factor. Therefore

$$\Omega(\rho_{2l} \otimes \mathbb{C}) = \bigcup_{j=1}^{l} \Omega(\psi_j \otimes \mathbb{C}) = \{\pm\theta_j, 1 \le j \le l\},$$

$$\chi_{\rho_{2l} \otimes \mathbb{C}}(\operatorname{Exp} H) = \sum_{j=1}^{l} (e^{2\pi i \theta_j(H)} + e^{-2\pi i \theta_j(H)}).$$

6. $G = SO(2l)$. Then $\operatorname{Ad} = \Lambda^2 \rho_{2l}$. Therefore

$$\operatorname{Ad} \otimes \mathbb{C}|T = \Lambda^2 \rho_{2l} \otimes \mathbb{C}|T = \Lambda^2(\rho_{2l} \otimes \mathbb{C}|T)$$

$$= \Lambda^2 \left(\sum_{j=1}^{l} \psi_j \otimes \mathbb{C} \right) = \Lambda^2 \left(\sum_{j=1}^{l} (\varphi_j \oplus \varphi_j') \right),$$

where $\Omega(\varphi_j) = \theta_j$, $\Omega(\varphi_j') = -\theta_j$. From here, it is straightforward to show that

$$\Delta(SO(2l)) = \{\pm\theta_j \pm \theta_k, j < k\}.$$

Exercise   To compute $\Delta(SO(2l+1))$.

3. Classification of Rank 1 Compact Connected Lie Groups

So far, we have already encountered three compact connected Lie groups of rank 1, namely, S^1, S^3 and SO(3). Are there any others? Let us try to find out.

Suppose G is such a Lie group, namely, compact connected and of rank 1. Let $\mathfrak{G}$ be its Lie algebra and $T \cong S^1$ be a maximal torus of G. Restricting the adjoint action of G on $\mathfrak{G}$ to T, one gets the decomposition

$$\begin{cases} \text{Ad}\,|T = 1 + \psi_1 + \cdots, \\ \mathfrak{G} = \mathbb{R}^1 \oplus \mathbb{R}^2(\psi_1) + \cdots, \end{cases}$$

where the non-trivial irreducible real representations $\{\psi_1, \ldots\}$ are all of the form

$$\psi_j : T \to \text{SO}(2); \quad e^{it} \mapsto \begin{pmatrix} \cos n_j t & -\sin n_j t \\ \sin n_j t & \cos n_j t \end{pmatrix}.$$

Of course, one may assume that

$$0 < n_1 \le n_2 \le \cdots \le n_j \le \cdots,$$

if there are more than one such ψ's.

Examples 1. In the case $G = S^3$, the above decomposition has only one two-dimensional one with $n_1 = 2$.

2. In the case $G = \text{SO}(3)$, the above decomposition has only one two-dimensional one with $n_1 = 1$.

Lemma $\mathfrak{G}_1 = \mathbb{R}^1 \oplus \mathbb{R}^2(\psi_1)$ *is a Lie subalgebra of* $\mathfrak{G}$ *and it is isomorphic to the Lie algebra of* S^3.

Proof: Let $H \in \mathbb{R}^1$ such that $\text{Exp}\,tH = e^{it} \in T \cong S^1$ and X_1, Y_1 be an orthonormal basis of $\mathbb{R}^2(\psi_1)$, namely

$$\begin{cases} \text{Ad}(\text{Exp}\,tH)X_1 = \cos n_1 t \cdot X_1 + \sin n_1 t \cdot Y_1, \\ \text{Ad}(\text{Exp}\,tH)Y_1 = -\sin n_1 t \cdot X_1 + \cos n_1 t \cdot Y_1. \end{cases}$$

Let us compute the bracket operations of H, X_1, Y_1 as follows. Differentiate the above equation with respect to t at $t = 0$, one gets

$$\begin{cases} [H, X_1] = \dfrac{d}{dt}\bigg|_{t=0} \mathrm{Ad}(\mathrm{Exp}\, tH) \cdot X_1 = n_1 Y_1, \\[3mm] [H, Y_1] = \dfrac{d}{dt}\bigg|_{t=0} \mathrm{Ad}(\mathrm{Exp}\, tH) \cdot Y_1 = -n_1 X_1. \end{cases}$$

Hence, by Jacobi identity, we have

$$[H, [X_1, Y_1]] = [[H, X_1], Y_1] + [X_1, [H, Y_1]] = 0.$$

Therefore, $[X_1, Y_1]$ must be a non-zero multiples of H, for otherwise, $\mathfrak{G}$ will contain a two-dimensional Abelian Lie subalgebra which clearly contradicts the assumption that G is of rank one! Set $[X_1, Y_1] = c \cdot H$. We shall show that $c > 0$. Recall that $\mathrm{Ad}(\mathrm{Exp}\, tX_1)$ are orthogonal transformation for all $t \in \mathbb{R}$. Hence

$$\langle \mathrm{Ad}(\mathrm{Exp}\, tX_1) \cdot Y_1, \mathrm{Ad}(\mathrm{Exp}\, tX_1) \cdot H \rangle \equiv \langle Y_1, H \rangle.$$

Differentiate the above equation at $t = 0$, one obtains

$$\langle [X_1, Y_1], H \rangle + \langle Y_1, [X_1, H] \rangle = 0,$$

which implies that

$$c \cdot \langle H, H \rangle = \langle [X_1, Y_1], H \rangle = -\langle Y_1, -n_1 Y_1 \rangle = n_1 \langle Y_1, Y_1 \rangle$$

$$\Rightarrow c = \frac{n_1 |Y_1|^2}{|H|^2} > 0.$$

From here, it is easy to show that $\{H, X_1, Y_1\}$ spans a Lie subalgebra isomorphic to the Lie algebra of S^3 (cf. Ex., Lecture 2). □

Theorem 3. *Let G be a compact connected Lie group of rank 1. Then G is isomorphic to one of the following examples, namely, S^1, S^3 or $SO(3)$.*

Proof: If G is commutative, then it is obvious that $G \cong S^1$. Let us assume that G is non-commutative. Then, by the lemma, its Lie algebra $\mathfrak{G}$ contains a Lie subalgebra $\mathfrak{G}_1$ isomorphic to that of S^3. Therefore, by Theorem 2.3 and the fact that S^3 is simply connected, there exists a compact connected Lie subgroup G_1, with $\mathfrak{G}_1$ as its Lie algebra, which is either isomorphic to S^3 or isomorphic to $SO(3)$. [In fact, $G_1 \cong S^3$ if $n_1 = 2$; $G_1 \cong SO(3)$ if $n_1 = 1$.] We shall show that $G = G_1$, namely, $\dim \mathfrak{G} = 3$.

Suppose the contrary that $\dim \mathfrak{G} > 3$, i.e. there are more than one two-dimensional irreducible components in the above decomposition of $\text{Ad}_T \mathfrak{G}$. Set

$$V = \sum_{j \geq 2} \mathbb{R}^2(\psi_j) = \mathfrak{G}_1^{\perp}, \qquad \varphi = (G_1, V).$$

Then

$$\Omega(\varphi \otimes \mathbb{C}) = \Omega(\varphi | T \otimes \mathbb{C})$$

$$= \Omega \left(\sum_{j \geq 2} \psi_j \otimes \mathbb{C} \right) = \bigcup_{j \geq 2} \{\Omega(\psi_j \otimes \mathbb{C})\}.$$

Recall that any complex irreducible representation of $SO(3)$ always contains one zero weight. Hence, the case $G \supsetneq G_1 \cong SO(3)$ is impossible because the above weight system contains no zero weight. Finally, the case $G \supsetneq G_1 \cong S^3$ is again impossible, because in this case, $n_1 = 2$ and

$$\Omega(\psi_j \otimes \mathbb{C}) = \{\pm n_j \theta\}, \quad n_j \geq n_1 = 2$$

$\Rightarrow \Omega(\psi \otimes \mathbb{C})$ contains no weight of the form $\pm \theta$ or 0 which is again a contradiction to Theorem 1.6 (cf. Example 1 in the above section). Hence G must be equal to G_1, namely, $G \cong S^3$ or $SO(3)$. □

Theorem 4.   *The multiplicity of every non-zero weight in $\Omega(\text{Ad}_G \oplus \mathbb{C})$ is always equal to 1, and moreover, for each root $\alpha \in \Delta(G)$, $k\alpha \in \Delta(G)$ if and only if $k = \pm 1$.*

Proof: Let $\mathfrak{G}$ be the Lie algebra of G, T be a maximal torus of G and $\mathfrak{h}$ be the Lie algebra of T. Then one has the following orthogonal decomposition of $\mathfrak{G}$ as Ad_T-invariant spaces

$$\mathfrak{G} = \mathfrak{h} \oplus \sum \mathbb{R}^2_{(\pm \alpha)},$$

where $\{\pm \alpha\}$ runs through pairs of non-zero weights in $\Omega(\text{Ad} \otimes \mathbb{C})$ with multiplicities. For $H \in \mathfrak{h}$, the action of $\text{Ad}(\text{Exp}\, H)$ on $\mathbb{R}^2_{(\pm \alpha)}$ is given by

$$\begin{pmatrix} \cos 2\pi\alpha(H) & -\sin 2\pi\alpha(H) \\ \sin 2\pi\alpha(H) & \cos 2\pi\alpha(H) \end{pmatrix}.$$

Let $\mathfrak{h}_\alpha$ be the kernel of $\alpha : \mathfrak{h} \to \mathbb{R}^1$, T_α be the subtorus of T with $\mathfrak{h}_\alpha$ as its Lie algebra, $G_\alpha = Z_G^0(T_\alpha)$ be the connected centralizer of T_α and $\tilde{G}_\alpha = G_\alpha / T_\alpha$. [In fact, Corollary 2 of Theorem 1 already proves that $Z_G(T_\alpha)$

is automatically connected; it is, however, not needed in this proof.] Let $\mathfrak{G}_\alpha$ be the Lie algebra of G_α. Then

$$\mathfrak{G}_\alpha = F(T_\alpha, \mathfrak{G}) = \mathfrak{h} \oplus \sum \mathbb{R}^2_{(\pm\beta)},$$

where $\{\pm\beta\}$ runs through those pairs of non-zero weights in $\Omega(\mathrm{Ad} \otimes \mathbb{C})$ with $\mathfrak{h}_\beta = \mathfrak{h}_\alpha$, namely, proportionate to α. Hence

$$\tilde{\mathfrak{G}}_\alpha \cong \mathfrak{G}_\alpha/\mathfrak{h}_\alpha = \mathfrak{h}/\mathfrak{h}_\alpha \oplus \sum \mathbb{R}^2_{(\pm\beta)},$$

and $T/T_\alpha \cong S^1$ is a maximal torus of $\tilde{G}_\alpha$, namely, $\tilde{G}_\alpha$ is a rank 1 compact connected Lie group. Thus, it follows from Theorem 3 that $\mathbb{R}^2_{(\pm\alpha)}$ is, in fact, the only component in the above direct sum. $\qquad\square$

Remarks   (i) From now on, the root system $\Delta(G)$ is proved to be a set with uniform multiplicity of 1.

(ii) The usual Cartan decomposition is exactly the complexification of the above decomposition of $\mathfrak{G}$, namely,

$$\mathfrak{G} \otimes \mathbb{C} = \mathfrak{h} \otimes \mathbb{C} \oplus \sum_{\alpha \in \Delta(G)} \mathbb{C}_\alpha, \quad \mathrm{Ad}(\mathrm{Exp}\,H)X_\alpha = e^{2\pi i\alpha(H)} \cdot X_\alpha,$$

for $H \in \mathfrak{h}$ and $X_\alpha \in \mathbb{C}_\alpha$. If one substitutes H by tH and then differentiates the above equation at $t = 0$, one gets

$$[H, X_\alpha] = 2\pi i\alpha(H) \cdot X_\alpha.$$

(iii) In the original decomposition of $\mathfrak{G}$ over the real, one has

$$\mathfrak{G} = \mathfrak{h} \oplus \sum \mathbb{R}^2_{(\pm\alpha)},$$

and

$$[H, Y_\alpha] = 2\pi\alpha(H) \cdot Z_\alpha,$$

$$[H, Z_\alpha] = -2\pi\alpha(H) \cdot Y_\alpha,$$

where $\{Y_\alpha, Z_\alpha\}$ is an orthonormal basis of $\mathbb{R}^2_{(\pm\alpha)}$.

Lecture 4

Coxeter Groups, Weyl Reduction
and Weyl Formulas

In this lecture, we shall continue the study of the orbital geometry of the adjoint transformation of G on both the manifold G and its Lie algebra $\mathfrak{G}$. Based upon the maximal tori theorem of É. Cartan and the Weyl reduction, i.e. $G/\mathrm{Ad} \cong T/W$ and $\mathfrak{G}/\mathrm{Ad} \cong \mathfrak{h}/W$, it is rather natural to consider the Weyl transformation groups (W, T) and $(W, \mathfrak{h})$ as the "vital core" of the geometry of non-commutativity of G. On the one hand, they are far-reaching simplifications of the adjoint actions of G on both G and $\mathfrak{G}$, and yet on the other hand, they retain the vital point of the orbit structures of the original adjoint transformations which is actually the geometric version of the totality of the non-commutativity of G. It is an added blessing that $(W, \mathfrak{h})$ are *generated by reflections*, namely, Coxeter groups. This naturally makes the basic geometry of Coxeter groups to become an important component of the structural theory of Lie groups.

1. Geometry of Coxeter Groups

Definition   A reflection is a differentiable involution $r : M \to M$ on a connected manifold M such that its fixed point set $F(r)$ is a codimension one submanifold which separates M into two connected regions interchanged by r.

Definition   A finite differentiable transformation group $W \times M \to M$ is called a group generated by reflections, or simply a Coxeter group, if W is generated by a collection of reflections.

Examples   1. Let S_n be the symmetric group of n letters and it acts on $\mathbb{R}^n = \{(x_1, x_2, \ldots, x_n); \, x_j \in \mathbb{R}\}$ by permuting the coordinates. Then it is a Coxeter group (generated by those reflections which are exactly those transpositions).

2. Let $\mathbb{R}^{n-1}$ be the subspace in the above $\mathbb{R}^n$ which is defined by $\sum x_j = 0$. Then it is an invariant subspace of the above S_n-action and $(S_n, \mathbb{R}^{n-1})$ is again a Coxeter group.

3. If the angle between two intersecting lines l_1, l_2 is π/n, then the subgroup of isometries generated by the two reflections with respect to l_1, l_2 is a group of order $2n$. It is one of the simplest example of Coxeter group.

4. Let $G = U(n)$, T be the subgroup of diagonal matrices and $\mathfrak{h}$ be the Lie algebra of T. Then T is a maximal torus of G and $\mathfrak{h}$ is the vector space of diagonal skew hermitian matrices, namely,

$$\mathfrak{h} = \left\{ \begin{pmatrix} i\theta_1 & & & \\ & i\theta_2 & & \\ & & \ddots & \\ & & & i\theta_n \end{pmatrix} ; \theta_j \in \mathbb{R} \right\} \cong \mathbb{R}^n.$$

Let $g \in N_G(T)$ be an arbitrary element of $N_G(T)$, i.e. $g^{-1}Tg = T$. Then $g(e_j)$, $1 \leq j \leq n$, are again eigenvectors of all elements of T and hence are multiples of a suitable e_l, namely,

$$g(e_j) = \lambda_j e_{l_j}, \quad 1 \leq j \leq n$$

where $|\lambda_j| = 1$ and $(l_1, l_2, \ldots, l_n)$ is a permutation of $(1, 2, \ldots, n)$. From here, it is not difficult to see that the Weyl transformation group $(W, \mathfrak{h})$ is, in fact, isomorphic to the above example $(S_n, \mathbb{R}^n)$.

5. Let $G = \mathrm{SU}(n)$, T be the subgroup of diagonal matrices and $\mathfrak{h}$ be the Lie algebra of T. The T is a maximal torus of G and

$$\mathfrak{h} = \left\{ \begin{pmatrix} i\theta_1 & & & \\ & i\theta_2 & & \\ & & \ddots & \\ & & & i\theta_n \end{pmatrix} ; \ \theta_j \in \mathbb{R}, \sum \theta_j = 0 \right\} \cong \mathbb{R}^{n-1}.$$

In this case, the Weyl transformation group $(W, \mathfrak{h})$ is isomorphic to the $(S_n, \mathbb{R}^{n-1})$ of Example 2.

6. Let (W, M) be a Coxeter group, $r \in W$ be a reflection and $\sigma \in W$ be an arbitrary element of W. Then $\sigma r \sigma^{-1}$ is also a reflection and $F(\sigma r \sigma^{-1}) = \sigma F(r)$.

Lemma 1. *Let $\tilde{\Delta}$ be the set of all reflections in a Coxeter group (W, M). Then W acts transitively on the set of connected components of $M \backslash \cup \{F(r); r \in \tilde{\Delta}\}$.*

Proof: Since $\sigma \tilde{\Delta} \sigma^{-1} = \tilde{\Delta}$ and $F(\sigma r \sigma^{-1}) = \sigma F(r)$, it is clear that

$$\bigcup \{F(r); r \in \tilde{\Delta}\} \quad \text{and} \quad M \backslash \bigcup \{F(r); r \in \tilde{\Delta}\}$$

are both invariant subsets of W. Therefore, the connected components of $M \backslash \bigcup \{F(r); r \in \tilde{\Delta}\}$ are permuted among themselves under the action of W. We shall call the above components chambers and prove the transitivity of the above W-action on the set of all chambers.

Observe that if C, C' are two chambers separated by a wall supported by $F(r)$, namely,

$$\dim \tilde{C} \cap F(r) \cap \tilde{C}' = \dim F(r),$$

then $r(C) = C'$. Therefore, if $\{C_0, C_1, \ldots, C_l\}$ is a sequence of chambers such that each consecutive pair $\{C_i, C_{i+1}\}$ are separated by a wall, say on $F(r_i)$, then

$$C_{i+1} = r_i(C_i) \quad \text{and} \quad C_l = r_{l-1} \cdot r_{l-2} \cdots r_0(C_0).$$

Hence, what one needs to show is that any two chambers can be linked by a sequence of chambers with common walls between consecutive ones, where such sequences are called *chains*.

For a pair of distinct reflections r, $r' \in \tilde{\Delta}$, it is clear that

$$\dim F(r) \cap F(r') \le \dim M - 2,$$

and hence, the union of all such subsets

$$\sum = \bigcup \{F(r) \cap F(r'); r \ne r' \in \tilde{\Delta}\}$$

is of *codimension* $= 2$. Therefore, $M \backslash \sum$ is still connected because a subset of codimension > 1 cannot separate a connected manifold even locally. This means that one can always go from one chamber to any other chamber by a pathway which only crosses the common walls between two consecutive chambers. This shows that any two chambers can be connected by a chain of chambers and hence the W-action on chambers is transitive.   □

Remarks   (i) The transitivity of the W-action on the set of chambers shows that all chambers are of equal standing. Hence it is convenient to fix one of them as the base chamber. We shall denote it by C_0 and call it the (chosen) *Weyl chamber* of (W, M).

(ii) To each $r \in \tilde{\Delta}$, $M \backslash F(r)$ consists of two connected components. We shall denote the "half-space" containing the above C_0 by M_r^+ and the other one by M_r^-. $M_r^\pm$ are respectively called the positive and negative half-space of the reflection r. It is not difficult to see that

$$C_0 = \bigcap \{M_r^+; r \in \tilde{\Delta}\}.$$

Definition   Let π be the subset of reflections in $\tilde{\Delta}$ whose fixed point set contains a wall of $\bar{C}_0$, namely,

$$\dim F(r) \cap \tilde{C}_0 = \dim M - 1.$$

Lemma 2. π *also forms a generator system of* W.

Proof: Let W' be the subgroup of W generated by π. We shall show that $W' \supset \tilde{\Delta}$ and hence $W' = W$.

Let $C_0, C_1, \ldots, C_l$ be a chain and $F(r)$ contains a wall of C_l. We shall prove by induction on l that $r \in W'$. Let r' be the reflection such that $F(r')$ contains the wall between C_{l-1} and C_l. Then, by the induction assumption,

$r' \in W'$. Since $r'(C_l) = C_{l-1}$, $F(r'rr') = r'(F(r))$ contains a wall of $r'(C_l) = C_{l-1}$. Again, by the induction assumption, $r'rr'$ also belongs to W'. Hence $r \in W'$. □

Definition π is called a simple system of generators of W, its elements will be henceforth denoted by $\{r_i; 1 \leq i \leq k\}$ and called the simple generators of W. To each $\sigma \in W$, $l(\sigma)$ is defined to be the *minimal* length of expressing σ as a product of the simple generators.

Lemma 3.   *Let $\sigma = r_{i_1} \cdot r_{i_2} \cdots r_{i_l}$, $l = l(\sigma)$, be a given expression of σ of minimal length. Set*

$$\sigma_j = r_{i_1} \cdot r_{i_2} \cdots r_{i_j}, \quad F_j = \sigma_{j-1} F(r_{i_j}) = F(\sigma_{j-1} r_{i_j} \sigma_{j-1}^{-1}),$$

and

$$C_j = \sigma_j(C_0), \quad 0 \leq j \leq l.$$

Then

(i) $C_0, C_1, \ldots, C_j, \ldots, C_l = \sigma(C_0)$ *is a shortest chain linking C_0 to $\sigma(C_0)$,*

(ii) *the set of hyperplanes $\{F_j, 1 \leq j \leq l\}$ is exactly the set of those hyperplanes separating C_0 and $\sigma(C_0)$ and hence it only depends on σ.*

Proof: By definition, $\bar{C}_0 \cap F(r_i)$ is a wall of $\bar{C}_0$. Hence

$$\sigma_{j-1}(\bar{C}_0 \cap F(r_{i_j})) = \bar{C}_{j-1} \cap F_j, \quad F_j = F(\sigma_j \sigma_{j-1}^{-1})$$

is a common wall between C_{j-1} and $C_j = \sigma_j(C_0) = \sigma_j \sigma_{j-1}^{-1}(C_{j-1})$ and thus $\{C_0, C_1, \ldots, C_j, \ldots, C_l = \sigma(C_0)\}$ is a chain. In fact, it is not difficult to see that above construction establishes a bijective correspondence between the expressions of σ in terms of the simple generators and the set of chains linking C_0 to $\sigma(C_0)$. Therefore, a given expression is one of the shortest if and only if the corresponding chain is a shortest one linking C_0 to $\sigma(C_0)$.

Let $F(r)$ be a "hyperplane" that separates C_0 and $\sigma(C_0)$. Then any chain linking C_0 and $\sigma(C_0)$ must cross it at least once, namely, at least one of the common walls between consecutive chambers of the chain is contained in $F(r)$.

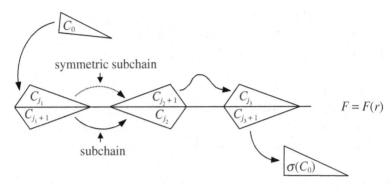

Fig. 2.

On the other hand, we claim that a shortest chain from C_0 to $\sigma(C_0)$ can only cross each separating hyperplane between C_0, $\sigma(C_0)$ exactly once! For otherwise, the chain can be shortened as indicated in Fig. 2, namely, simply by replacing the subchain indicated above by its symmetric sub-chain, one obtains a chain with two less chambers. This proves the second assertion.   □

Theorem 1.   *Let (W, M) be a group generated by reflections on M. Then W acts simply, transitively on the set of chambers and the closure of a chamber, say $\bar{C}_0$, forms a fundamental domain, i.e. $\bar{C}_0$ intersects every W-orbit exactly once.*

Proof: By Lemma 1, W acts transitively on the set of chambers. There-fore, what remains to be shown is that $\sigma(C_0) = C_0$ implies that $\sigma = \mathrm{Id}$. This follows easily from Lemma 3 and the fact that C_0 is, of course, a chain of zero length linking C_0 to $\sigma(C_0) = C_0$. Thus, by Lemma 3, $l(\sigma) = 0$ and $\sigma = \mathrm{Id}$.

Next suppose x_0 and $\sigma(x_0)$ both belong to $\bar{C}_0$. Then $\sigma(x_0) \in \bar{C}_0 \cap \sigma(\bar{C}_0) \Rightarrow C_0 \cup \{\sigma(x_0)\} \cup \sigma(C_0)$ is *connected*. Hence, every hyperplane sepa-rating C_0 and $\sigma(C_0)$ must cut through $\sigma(x_0)$. Therefore, by Lemma 3, $\sigma(x_0)$ is, in fact, fixed under σ, i.e. $\sigma^2(x_0) = \sigma(x_0)$, and hence $x_0 = \sigma^{-1}\sigma^2(x_0) = \sigma(x_0)$. This proves that every W-orbit can intersect $\bar{C}_0$ at most in one point. On the other hand,

$$W \cdot C_0 = M \backslash \bigcup \{F(r); r \in \tilde{\Delta}\} \Rightarrow W \cdot \bar{C}_0 = M,$$

namely, $\bar{C}_0$ intersects every W-orbit at least once. This shows that $\bar{C}_0$ intersects every W-orbit exactly once, namely, $\bar{C}_0$ is a fundamental domain of (W, M). □

Corollary 1. *The isotropy subgroup of a point x_0, W_{x_0}, is exactly the subgroup generated by those reflections whose fixed point set contains x_0.*

2. Geometry of $(W, \mathfrak{h})$ and the Root System

It is natural and convenient to equip the Lie algebra $\mathfrak{G}$ of a compact Lie group G with an Ad-invariant inner product. Thus, the restriction to $(W, \mathfrak{h})$ is a finite group of isometric transformations.

Theorem 2. *The Weyl transformation group $(W, \mathfrak{h})$ is a Coxeter group generated by those reflections $\{r_\alpha; \pm\alpha \in \Delta(G)\}$ where r_α is the reflection with respect to the hyperplane $\mathfrak{h}_\alpha = \ker\alpha$.*

Proof: To each pair of roots $\{\pm\alpha\}$, one has the Lie subgroup G_α whose Lie algebra

$$\mathfrak{G}_\alpha = \mathfrak{h} \oplus \mathbb{R}^2_{(\pm\alpha)} = \mathfrak{h}_\alpha \oplus \{\mathbb{R}^1 \oplus \mathbb{R}^2_{(\pm\alpha)}\} = \mathfrak{h}_\alpha \oplus \tilde{\mathfrak{G}}_\alpha,$$

where $\mathbb{R}^1$ is the perpendicular line of $\mathfrak{h}_\alpha$ in $\mathfrak{h}$ and $\tilde{\mathfrak{G}}_\alpha$ is isomorphic to the Lie algebra of S^3. Let $\tilde{G}_\alpha$ (resp. T_α) be the Lie subgroup with $\tilde{\mathfrak{G}}_\alpha$ (resp. $\mathfrak{h}_\alpha$) as its Lie algebra and $f : S^3 \to \tilde{G}_\alpha$ be the covering homomorphism. Then the following composition

$$T_\alpha \times S^3 \overset{s \times f}{\longrightarrow} G_\alpha \times G_\alpha \overset{m}{\longrightarrow} G_\alpha$$

is a covering homomorphism. Therefore, the Weyl group of G_α and that of $T_\alpha \times S^3$ are identical transformation groups on $\mathfrak{h}$, namely, $W(G_\alpha) \simeq \mathbb{Z}_2$ and acts on $\mathfrak{h}$ as the reflection with respect to the hyperplane $\mathfrak{h}_\alpha$.

Let W' be the subgroup in W generated by the collection of reflections $\{r_\alpha \in W(G_\alpha); \pm\alpha \in \Delta(G)\}$. Since the root system $\Delta(G) \subset \mathfrak{h}^*$ is clearly an invariant subset under the induced W-action on $\mathfrak{h}^*$, W' is a normal subgroup of W.

Let $\{C_i\}$ be the set of chambers of $(W', \mathfrak{h})$, namely, the connected components of $\mathfrak{h} \backslash \bigcup\{\mathfrak{h}_\alpha; \pm\alpha \in \Delta(G)\}$ and C_0 be a chosen Weyl chamber. Both W' and W act on the above set of chambers as permutation groups and, by Lemma 3, W' acts simply transitively. Let W_0 be the subgroup of W which leaves C_0 invariant. Then $W = W'$ if and only if W_0 is the trivial subgroup of identity. Suppose the contrary that W_0 is non-trivial. Recall that C_0 is an open convex subset of $\mathfrak{h}$, the center of mass of a W_0-orbit in C_0 is again in C_0, thus producing a fixed point, say X_0, in C_0. Therefore,

$$G_{X_0}^0 = T \quad \text{but} \quad G_{X_0}/T \supset W_0,$$

namely, G_{X_0} is disconnected. Let S be the closure of $\{\operatorname{Exp} tX_0; t \in \mathbb{R}\}$. Then S is a torus subgroup of G and $G_{X_0} = Z_G(S)$. Hence, by Corollary 2 of Theorem 3.1, $G_{X_0} = Z_G(S)$ is connected. The above contradiction proves that W_0 must be trivial and hence $W' = W$. $\qquad\square$

Next let us apply the results of Section 1 to the above special case of Weyl transformation group $(W, \mathfrak{h})$. Since $\mathfrak{h}$ has already been equipped with a W-invariant inner product, it is convenient to consider the root system $\Delta(G)$ as a subset of $\mathfrak{h}$ via the following identification, namely,

$$\iota : \mathfrak{h}^* \cong \mathfrak{h}, \quad \alpha(H) = (\iota(\alpha), H), H \in \mathfrak{h}.$$

In this setting, W is an orthogonal transformation group generated by the reflections with respect to roots, namely,

$$r_\alpha(H) = H - \frac{2(\alpha, H)}{(\alpha, \alpha)}\alpha, \pm\alpha \in \tilde{\Delta},$$

$$F(r_\alpha, \mathfrak{h}) = \langle \alpha \rangle^\perp = \{H \in \mathfrak{h}, (\alpha, H) = 0\}.$$

By choosing a Weyl chamber C_0, then a root $\alpha \in \Delta$ is said to be positive (resp. negative) if α and C_0 is at the same (resp. opposite) side of $\langle \alpha \rangle^\perp$, namely,

$$\alpha \in \Delta^+(\text{resp. } \Delta^-) \Leftrightarrow (\alpha, C_0) > 0 \ (\text{resp. } < 0).$$

Conversely, C_0 and $\bar{C}_0$ can also be characterized as follows:

$$C_0 = \{H \in \mathfrak{h}; (\alpha, H) > 0, \alpha \in \Delta^+\},$$

$$\bar{C}_0 = \{H \in \mathfrak{h}; (\alpha, H) \geq 0, \alpha \in \Delta^+\}.$$

Moreover, the *system of simple roots*, π, corresponding to the choice of C_0 is exactly the *minimal subset* of Δ^+ such that

$$C_0 = \{H \in \mathfrak{h}; (\alpha_i, H) > 0, \alpha_i \in \pi\}.$$

Geometrically, they are exactly those positive roots α_i such that $\langle \alpha_i \rangle^\perp$ contains a wall of $\bar{C}_0$. Algebraically, they are exactly the "indecomposable elements" of Δ^+, namely, those positive roots which cannot be decomposed into the sum of positive roots. For example, if $\alpha = \alpha_1 + \alpha_2, \alpha, \alpha_1, \alpha_2 \in \Delta^+$, then the condition $(\alpha, H) > 0$ is already implied by $(\alpha_1, H) > 0$ and $(\alpha_2, H) > 0$ and hence can be omitted from the defining condition of C_0. We shall prove later that the set of indecomposable elements of Δ^+ is linearly independent, (cf. the remark following Lemma 6).

Lemma 4. *Let $\Omega(\varphi)$ be the weight system of a complex representation, φ, of G. Then*

(i) $\frac{2(\omega,\alpha)}{(\alpha,\alpha)} \in \mathbb{Z}$ *for $\omega \in \Omega(\varphi)$ and $\alpha \in \Delta(G)$,*

(ii) $m(\omega, \varphi) \leq m(\omega - j\alpha, \varphi)$ *for all $0 \leq j \leq \frac{2(\omega,\alpha)}{(\alpha,\alpha)}$ or $\frac{2(\omega,\alpha)}{(\alpha,\alpha)} \leq j \leq 0$.*

Proof: The restriction of φ to G_α can be considered as a representation of $T_\alpha \times S^3$. Every complex *irreducible* representation of $T_\alpha \times S^3$ is an outer tensor product of a one-dimensional representation of T_α and an irreducible representation of S^3. Therefore, by Theorem 1.7, the weight system of any irreducible representation of G_α forms an α-*string* invariant under r_α. Therefore, the weight system $\Omega(\varphi)$ is a union of such r_α-symmetric α-string which clearly satisfies both (i) and (ii). $\qquad\square$

Lemma 5. *In the special case of $\varphi = \mathrm{Ad}_G \otimes \mathbb{C}$, one has*

(i) $\frac{2(\alpha,\beta)}{(\alpha,\alpha)} = -(p+q)$ *for $\alpha, \beta \in \Delta(G)$, where $\{\beta + j\alpha; p \geq j \geq q\}$ is the unique α-string containing β,*

(ii) $(\alpha_i, \alpha_j) \leq 0$ *for $\alpha_i \neq \alpha_j \in \pi$ (i.e., distinct simple roots).*

Proof: (i) Since the multiplicities of roots are always 1, there is a unique α-string passing through a given β. It follows from the r_α-invariance that

$$\beta + q\alpha = r_\alpha(\beta + p\alpha) = \beta + p\alpha - \frac{2(\alpha, \beta + p\alpha)}{(\alpha, \alpha)} \cdot \alpha$$

$$= \beta - \frac{2(\alpha, \beta)}{(\alpha, \alpha)}\alpha - p\alpha.$$

Hence

$$\frac{2(\alpha, \beta)}{(\alpha, \alpha)} = -(p + q).$$

(ii) Since simple roots are indecomposable, $\alpha_i - \alpha_j \notin \Delta$. For otherwise, either $\alpha_i - \alpha_j$ or $\alpha_j - \alpha_i$ is a positive root and hence, either α_i or α_j is decomposable. Therefore, $q = 0$ and

$$\frac{2(\alpha_i, \alpha_j)}{(\alpha_i, \alpha_i)} = -(p+q) = -p \leq 0. \qquad \square$$

Lemma 6.   *The system of simple roots, π, is a linearly independent set and the angle between a pair of simple roots is either $\pi/2, 2\pi/3, 3\pi/4$ or $5\pi/6$.*

Proof: (i) Suppose the contrary that there exists a non-trivial linear relation among the simple roots. Then the coefficients cannot be all of the same sign because $(\alpha_i, H) > 0$ for all $\alpha_i \in \pi$ and $H \in C_0$. Let the non-trivial linear relation be

$$\sum_{\alpha_i \in \pi'} \lambda_i \alpha_i - \sum_{\alpha_j \in \pi''} \mu_j \alpha_j = 0, \quad \lambda_i, \mu_j > 0.$$

Then

$$\left| \sum_{\alpha_i \in \pi'} \lambda_i \alpha_i \right|^2 = \left(\sum \lambda_i \alpha_i, \sum \mu_j \alpha_j \right)$$

$$= \sum \lambda_i \mu_j (\alpha_i, \alpha_j) \leq 0, \quad [\text{all } (\alpha_i, \alpha_j) \leq 0].$$

This is clearly a contradiction because $\sum \lambda_i \alpha_i$ is a linear combination of uniform positive coefficients and hence must be non-zero.

(ii) Since $2(\alpha_i, \alpha_j)/(\alpha_i, \alpha_i)$ are non-positive integers and

$$0 \leq \frac{2(\alpha_i, \alpha_j)}{(\alpha_i, \alpha_i)} \cdot \frac{2(\alpha_i, \alpha_j)}{(\alpha_j, \alpha_j)} \leq 3,$$

it is easy to see that there are only the following four cases:

$$\left\{ \frac{2(\alpha_i, \alpha_j)}{(\alpha_i, \alpha_i)}, \frac{2(\alpha_i, \alpha_j)}{(\alpha_j, \alpha_j)} \right\} = \begin{cases} \{0, 0\} \\ \{-1, -1\} \\ \{-1, -2\} \\ \{-1, -3\} \end{cases} ; \quad \text{angle} = \begin{cases} \dfrac{\pi}{2}, \\ \dfrac{2\pi}{3}, \\ \dfrac{3\pi}{4}, \\ \dfrac{5\pi}{6}. \end{cases} \qquad \square$$

Remark   Then above lemma still holds if one replace π by the subset of indecomposable elements in Δ^+.

Lemma 7.   (i) $r_i(\Delta^+) = (\Delta^+ \backslash \{\alpha_i\}) \cup \{-\alpha_i\}$,

(ii) *Set* $\delta = \frac{1}{2} \sum_{\alpha \in \Delta^+} \alpha$. *Then* $2(\alpha_i, \delta)/(\alpha_i, \alpha_i) = 1$ *for all* $\alpha_i \in \pi$.

Proof: Let β be an arbitrary element in $\Delta^+ \backslash \{\alpha_i\}$. Then C_0 and $r_i(C_0)$ are both at the positive side of $\langle \beta \rangle^\perp$ because $\langle \alpha_i \rangle^\perp$ is the only hyperplane which separates C_0 and $r_i(C_0)$. Hence

$$(\beta, r_i(C_0)) > 0 \Rightarrow (r_i(\beta), C_0) > 0 \Rightarrow r_i(\beta) \in \Delta^+,$$

namely, r_i *permutes* elements of $\Delta^+ \backslash \{\alpha_i\}$ and sends α_i to $-\alpha_i$. Therefore, $r_i(\Delta^+) = (\Delta^+ \backslash \{\alpha_i\}) \cup \{-\alpha_i\}$.

$$\delta - \frac{2(\alpha_i, \delta)}{(\alpha_i, \alpha_i)} \alpha_i = r_i(\delta) = \frac{1}{2} \sum_{\alpha \in \Delta^+} r_i(\alpha)$$

$$= \frac{1}{2} \sum_{\alpha \in \Delta^+} \alpha - \frac{1}{2}\alpha_i - \frac{1}{2}\alpha_i = \delta - \alpha_i,$$

and hence

$$\frac{2(\alpha_i, \delta)}{(\alpha_i, \alpha_i)} = 1 \quad \text{for all } \alpha_i \in \pi. \qquad \square$$

Remark   The above inner product implies that $\delta \in C_0$ and hence $W(\delta) = \{\sigma(\delta); \sigma \in W\}$ consists of $|W|$ distinct points.

Examples: Root Systems of Rank 2   The cardinal number of the system of simple roots $\pi \subset \Delta^+ \subset \Delta$ is defined to be the rank of a root system Δ. There are the following four possibilities for root system of rank 2, according to the angle between α_1, α_2 is $\pi/2$, $2\pi/3$,

$3\pi/4$ or $5\pi/6$:

(i) $\frac{\pi}{2}$-case:

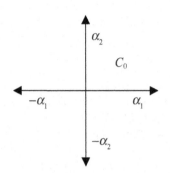

(ii) $\frac{2\pi}{3}$-case:

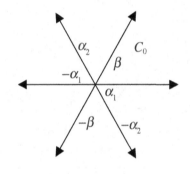

$$\beta = \alpha_1 + \alpha_2$$

(iii) $\frac{3\pi}{4}$-case:

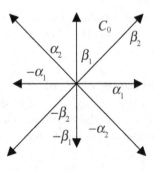

$$\beta_1 = \alpha_1 + \alpha_2, \beta_2 = 2\alpha_1 + \alpha_2$$

(iv) $\frac{5\pi}{6}$-case:

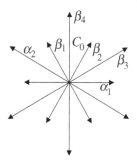

$\beta_1 = \alpha_2 + \alpha_1, \quad \beta_2 = \alpha_2 + 2\alpha_1, \quad \beta_3 = \alpha_2 + 3\alpha_1, \quad \beta_4 = 2\alpha_2 + 3\alpha_1.$

Exercises: 1. Compute the Weyl groups for each of the above cases.

2. Use Lemma 5 to show the above four cases are the only possible cases.

3. The Volume Function and Weyl Integral Formula

In applying the character theory to study the representations of a given compact group G, one needs to compute the hermitian products of character functions in $L_2(G)$ which are, by definition, integrals of the following form

$$\int_G \chi_\varphi(g) \cdot \overline{\chi_\psi(g)} dg$$

with respect to the Haar measure of total measure 1. In the nice situation of a compact connected Lie group G, one may equip G with a bi-invariant Riemannian metric with total volume 1 and effectively reduce the above integration of an Ad_G-*invariant function* over G to a much simpler integration of a W-*invariant function* over a maximal torus T.

Theorem 3. *Let $f(g)$ be an Ad_G-invariant function defined on G and $f(t)$ be its restriction to a given maximal torus T which is, of course,*

W-invariant. Then

$$\int_G f(g)dg = \frac{1}{|W|}\int_T f(t) \cdot v(t)dt,$$

where $|W|$ is the order of W and $v(t)$ is the m-dimensional volume of the orbit $G(t)$, $m = \dim G - \dim T$.

Proof: As has already been pointed out in Remark (2) following the proof of Theorem 1 of Lecture 3 (see page 45), the union of all orbits of dimensions lower than m is a subset of measure zero and hence can be omitted without affecting the values of the above integral.

Every m-dimensional G-orbit intersects T perpendicularly and transversally at exactly $|W|$-points and T is a totally geodesic submanifold of G. Since $f(g)$ is assumed to be Ad_G-invariant, i.e., constant along each orbit, it is convenient to integrate firstly along the orbital directions and then along the T-directions. Hence

$$\int_G f(g)dg = \int_{T/W} f(t)v(t)dt = \frac{1}{|W|}\int_T f(t)v(t)dt. \qquad \square$$

In order to fully exploit the above reduction formula of H. Weyl, one needs to compute a nice, explicit form of the above volume function. Every m-dimensional G-orbit is a *homogeneous* Riemannian manifolds of the same type of G/T. Let us take a fixed homogeneous metric on G/T with total volume 1. Then, to each given m-dimensional orbit $G(t)$, one has the following *equivariant* bijection:

$$G/T \xrightarrow{\quad B_t \quad} G(t)$$

$$g \cdot T \longmapsto g(t) = gtg^{-1}.$$

Notice that all the tangent spaces at points in both G/T and $G(t)$ are already equipped with inner products, (i.e. the Riemannian structures on both G/T and $G(t)$), and the Jacobian, i.e. $\det(dB_t|_x)$, $x \in G/T$, records the magnification factor of the volume element at x. Since B_t is an equivariant map between homogeneous Riemannian manifolds, the Jacobian function of B_t:

$$J(t) = \det(dB_t|_x), \quad x \in G/T$$

is a *constant function*, namely, B_t is a map of uniform magnification. Hence,

$$v(t) = \mathrm{vol}_m(G(t)) : \mathrm{vol}_m(G/T) = \det(dB_t|_{x_0}),$$

where x_0 (= the coset $e \cdot T$) is the base point of G/T. This enables us to reduce the computation of $v(t)$ to that of $\det(dB_t|_{x_0})$.

To each pair of roots $\{\pm\alpha\}$, one has the subgroup G_α. Notice that G_α/T is a round two-sphere imbedded in G/T, say denoted by S_α^2, and its tangent space at the base point x_0 is exactly the T-irreducible subspace $\mathbb{R}^2_{(\pm\alpha)}$ in $T_{x_0}(G/T)$. Therefore, one has the following commutative diagrams of maps, namely,

$$G/T \xrightarrow{\quad B_t \quad} G(t)$$

$$\cup \qquad\qquad \cup$$

$$S_\alpha^2 = G_\alpha/T \xrightarrow{\quad B_t^\alpha \quad} G_\alpha(t) = S_\alpha^2(t),$$

and its linearization at x_0

$$T_{x_0}(G/T) = \oplus \sum T_{x_0}(S_\alpha^2) \xrightarrow{\quad dB_t \quad} \oplus \sum T_t S_\alpha^2(t) = T_t G(t)$$

$$\cup \qquad\qquad\qquad\qquad \cup$$

$$T_{x_0}(S_\alpha^2) \xrightarrow{\quad dB_t^\alpha \quad} T_t S_\alpha^2(t).$$

Hence

$$v(t) = \det(dB_t) = \prod_{\alpha \in \Delta^+} \det(dB_t^\alpha).$$

Notice that the geometric meaning of each factor is that

$$\det(dB_t^\alpha) = \mathrm{Area}(S_\alpha^2(t)) : \mathrm{Area}(S_\alpha^2),$$

where $\mathrm{Area}(S_\alpha^2)$ is a constant and $S_\alpha^2(t)$ are conjugacy classes in G_α. Set $t = \mathrm{Exp}\, H$, $H \in \mathfrak{h}$, and $T_\alpha \times S^3 \to G_\alpha$ be the covering homomorphism. Then the area of $S_\alpha^2(\mathrm{Exp}\, H)$ in G_α and that of its inverse image are only differed by a constant. Therefore

$$\mathrm{Area}(S_\alpha^2(\mathrm{Exp}\, H)) = c \sin^2 \pi\alpha(H) \quad \text{or} \quad c \cdot \sin^2 \pi(\alpha, H).$$

[Recall that the root of S^3 is $\pm 2\theta$ and the area of $S^2(\mathrm{Exp}\, H)$ in S^3 is $4\pi \sin^2 2\pi\theta(H)$.]

Theorem 3'.   *Let $v(\mathrm{Exp}\,H)$, $H \in \mathfrak{h}$, be the m-dimensional volume of the conjugacy class of $\mathrm{Exp}\,H$, $m = \dim G - \mathrm{rk}(G)$, as a submanifold in G with a bi-invariant Riemannian metric of total volume 1. Then*

$$v(\mathrm{Exp}\,H) = \prod_{\alpha \in \Delta^+} (4\sin^2 \pi(\alpha, H))$$

$$= Q(H) \cdot \overline{Q(H)},$$

$$Q(H) = \sum_{\sigma \in W} \mathrm{sign}(\sigma) e^{2\pi i(\sigma \cdot \delta, H)}, \quad \delta = \frac{1}{2} \sum_{\alpha \in \Delta^+} \alpha.$$

Proof: It follows from the above discussion that

$$v(\mathrm{Exp}\,H) = c \cdot \prod_{\alpha \in \Delta^+} (4\sin^2 \pi(\alpha, H))$$

$$= c \cdot \prod_{\alpha \in \Delta^+} |e^{\pi i(\alpha, H)} - e^{-\pi i(\alpha, H)}|^2$$

for a suitable constant c. Set

$$Q(H) = \prod_{\alpha \in \Delta^+} (e^{\pi i(\alpha, H)} - e^{-\pi i(\alpha, H)}).$$

We shall prove that $c = 1$ and

$$Q(H) = \sum_{\sigma \in W} \mathrm{sign}(\sigma) e^{2\pi i(\sigma \cdot \delta, H)}.$$

Let us first establish the above identity, where $\mathrm{sign}(\sigma) = (-1)^{l(\sigma)}$. Observe that, for each reflection r_j with respect to the simple root α_j,

$$Q(r_j H) = \prod_{\alpha \in \Delta^+} (e^{\pi i(\alpha, r_j H)} - e^{-\pi i(\alpha, r_j H)})$$

$$= \prod_{\alpha \in \Delta^+} (e^{\pi i(r_j \alpha, H)} - e^{-\pi i(r_j \alpha, H)}),$$

and it follows from Lemma 7, i.e. $r_j(\Delta^+) = (\Delta^+ \backslash \{\alpha_j\}) \cup \{-\alpha_j\}$, that the actual difference between the above product and the original product is that

(i) $(e^{\pi i(\alpha_j, H)} - e^{-\pi i(\alpha_j, H)})$ is replaced by its negative,
(ii) other factors are permuted in their ordering.
    Therefore

$$Q(r_j H) = -Q(H),$$

and hence

$$Q(\sigma H) = \text{sign}(\sigma)Q(H),$$

namely, $Q(H)$ is an alternating function with respect to the W-action on $\mathfrak{h}$. In expanding the product form of $Q(H)$, the leading term is

$$\prod_{\alpha \in \Delta^+} e^{\pi i(\sigma, H)} = e^{\pi i(2\delta, H)} = e^{2\pi i(\delta, H)}.$$

Hence, by the alternating property of $Q(H)$,

$$Q(H) = \sum_{\sigma \in W} \text{sign}(\sigma)e^{2\pi i(\sigma\delta, H)} + \text{possible other terms}.$$

However, the fact

$$\frac{2(\delta, \alpha_i)}{(\alpha_i, \alpha_i)} = 1 \quad \text{for all } \alpha_i \in \pi$$

show that δ is the only vector of the forms

$$\frac{1}{2}\sum_{\alpha \in \Delta^+} \pm\alpha \quad \text{(with all possible choices of signs)}$$

which belongs to C_0. Therefore, there is, in fact, *no other terms*.

Finally, let us show that $c = 1$. Substitute $c \cdot Q(H) \cdot \overline{Q(H)}$ for $v(\text{Exp } H)$ into the formula of Theorem 3 with $f \equiv 1$, one gets

$$1 = \int_G 1 \cdot dg = \frac{1}{|W|}\int_T c \cdot Q(H) \cdot \bar{Q}(H)dt$$

$$= \frac{c}{|W|} \cdot \left| \sum_{\sigma \in W} \text{sign}(\sigma)e^{2\pi i(\sigma\delta, H)} \right|^2_{L_2(T)}.$$

Notice that $\{e^{2\pi i(\sigma\delta, H)}, \sigma \in W\}$ is a set of $|W|$ orthonormal vectors in $L_2(T)$. Hence

$$\left| \sum_{\sigma \in W} \text{sign}(\sigma)e^{2\pi i(\sigma\delta, H)} \right|^2 = |W| \Rightarrow c = 1. \qquad \square$$

4. Weyl Character Formula and Classification of Complex Irreducible Representations

Let φ be a complex irreducible representation of G, T be a maximal torus with $\mathfrak{h}$ as its Lie algebra. Let $\Omega(\varphi)$ be the weight system of φ and χ_φ be its character function. Then

$$\chi_\varphi(\operatorname{Exp} H) = \sum_{\omega \in \Omega(\varphi)} m(\omega, \varphi) e^{2\pi i(\omega, H)}, \quad H \in \mathfrak{h},$$

and it is a W-invariant function. We shall apply the integration formula of Theorem 3 in Sec. 3 to compute the following integral criterion of irreducibility, namely,

$$1 = \int_G \chi_\varphi \cdot \bar{\chi}_\varphi dg = \frac{1}{|W|} \int_T \chi_\varphi(\operatorname{Exp} H) \cdot \bar{\chi}_\varphi(\operatorname{Exp} H) \cdot Q(H) \cdot \bar{Q}(H) dt$$

$$= \frac{1}{|W|} |\chi_\varphi(\operatorname{Exp} H) \cdot Q(H)|^2_{L_2(T)}.$$

Since $\chi_\varphi(\operatorname{Exp} H)$ is W-invariant and $Q(H)$ is W-alternating, it is clear that $\chi_\varphi(\operatorname{Exp} H) \cdot Q(H)$ is a W-*alternating function*.

Set

$$\sigma \cdot f(t) = f(\sigma^{-1} t), \quad \sigma \in W, \quad t \in T, \quad f \in L_2(T).$$

Then

$$P = \frac{1}{|W|} \sum_{\sigma \in W} \operatorname{sign}(\sigma) \sigma : L_2(T) \to L_2(T)$$

is an orthogonal projection of $L_2(T)$ onto the subspace of W-alternating L_2-functions. [It is easy to verify that $P^2 = P$ and for any $f \in L_2(T)$, Pf is W-alternating.]

Let Γ be the set of all $\omega \in \mathfrak{h}$ with

$$\frac{2(\omega, \alpha_i)}{(\alpha_i, \alpha_i)} \in \mathbb{Z}$$

and $\dot{\Gamma}_0 = \Gamma \cap C_0$. Then

$$\{e^{2\pi i(\omega, H)}; \omega \in \Gamma\}$$

forms an orthonormal basis of $L_2(T)$ and it is not difficult to see that

$$\left\{ \sqrt{|W|} \cdot P e^{2\pi i(\omega, H)}, \omega \in \Gamma_0 \right\}$$

forms an orthonormal basis of the subspace of alternating L_2-functions. [Notice that $|P \cdot e^{2\pi i(\omega, H)}|^2 = 1/|W|$ for $\omega \in \Gamma_0$.]

For the following discussion, it is convenient to introduce an ordering on $\mathfrak{h}$ as follows.

Definition   Fix an ordering of the simple roots and then extend them to a basis of $\mathfrak{h}$ by adding vectors if necessary. An element of $\mathfrak{h}$ is defined to be positive if its *first non-zero coordinate* with respect to the above ordered basis is positive.

Remark   The above ordering is clearly rather arbitrarily fixed. It depends on the choice of C_0 and the ordering of simple roots. Anyhow, it will only serve the limited purpose of providing some convenience in book-keeping.

Definition   The highest element in $\Omega(\varphi)$ is called the highest weight of φ, and shall be denoted by Λ_φ.

Theorem 4.   (i) *The multiplicity of the highest weight of a complex irreducible representation φ is always 1.*

(ii) *Two complex irreducible representations, φ and ψ, of G are equivalent if and only if $\Lambda_\varphi = \Lambda_\psi$.*

(iii)

$$\chi_\psi(\operatorname{Exp} H) = \frac{\sum_{\sigma \in W} \operatorname{sign} \sigma\, e^{2\pi i(\sigma(\Lambda_\varphi + \delta), H)}}{\sum_{\sigma \in W} \operatorname{sign} \sigma\, e^{2\pi i(\sigma\delta, H)}}.$$

Proof: Let m_0 be the multiplicity of the highest weight, Λ_φ, in $\Omega(\varphi)$. Then

$$\chi_\varphi(\operatorname{Exp} H) = m_0 e^{2\pi i(\Lambda_\varphi, H)} + \text{terms of lower order},$$

$$Q(H) = e^{2\pi i(\delta, H)} \pm \text{terms of lower order}.$$

Hence

$$\chi_\varphi(\operatorname{Exp} H) \cdot Q(H) = m_0 \cdot e^{2\pi i(\Lambda_\varphi + \delta, H)} \pm \text{terms of lower order}.$$

Therefore, by its alternating property,

$$\chi_\varphi(\mathrm{Exp}\,H) \cdot Q(H) = m_0 \cdot \sum_{\sigma \in W} \mathrm{sign}\,\sigma e^{2\pi i(\sigma(\Lambda_\varphi + \delta), H)}$$

$$+ \text{ possible other alternating sums.}$$

However, it follows from the integral criterion of the irreducibility of φ that

$$|W| = |\chi_\varphi(\mathrm{Exp}\,H) \cdot Q(H)|^2_{L_2(T)} = m_0^2 \cdot |W| + |\text{possible terms}|^2.$$

Hence, the only possibility is that $m_0 = 1$ and

$$\chi_\varphi(\mathrm{Exp}\,H) \cdot Q(H) = m_0 \cdot \sum_{\sigma \in W} \mathrm{sign}\,\sigma e^{2\pi i(\sigma(\Lambda_\varphi + \delta), H)},$$

namely,

$$\chi_\psi(\mathrm{Exp}\,H) = \frac{\sum_{\sigma \in W} \mathrm{sign}\,\sigma e^{2\pi i(\sigma(\Lambda_\varphi + \delta), H)}}{\sum_{\sigma \in W} \mathrm{sign}\,\sigma e^{2\pi i(\sigma\delta, H)}}.$$

Since the character function $\chi_\varphi(\mathrm{Exp}\,H)$ is a complete invariant of φ, the second assertion follows readily from the above character formula of expressing $\chi_\varphi(\mathrm{Exp}\,H)$ purely in terms of its highest weight Λ_φ.   □

Corollary

$$\dim \varphi = \prod_{\alpha \in \Delta^+} \frac{(\Lambda_\varphi + \delta, \alpha)}{(\delta, \alpha)}.$$

Proof: The value of χ_φ at the identity e is, of course, just $\dim \varphi$. Therefore, one expects to obtain $\dim \varphi$ simply by substituting $H = 0$ into the above formula. However, such a substitution makes the above formula into an indeterminant form of $\frac{0}{0}$! Of course, that does not mean that the above formula cannot be suitably exploited to give us $\dim \varphi$. A typical way to get around such indeterminant forms is to find the limit of the quotient as $H \to 0$. As it turns out, the best way is to set $H = t \cdot \delta$ and then compute the limit of quotient as $t \to 0$, because one can again make use of the identity of (iii) as follows,

$$\sum_{\sigma \in W} \operatorname{sign} \sigma e^{2\pi i (\sigma(\delta), t\delta)} = \prod_{\alpha \in \Delta^+} \left(e^{\pi i (\alpha, \delta) t} - e^{-\pi i (\alpha, \delta) t} \right),$$

$$\sum_{\sigma \in W} \operatorname{sign} \sigma e^{2\pi i (\sigma(\Lambda_\varphi + \delta), t\delta)} = \sum_{\sigma \in W} \operatorname{sign} \sigma e^{2\pi i (\sigma(\delta), t(\Lambda_\varphi + \delta))}$$

$$= \prod_{\alpha \in \Delta^+} \left(e^{\pi i (\alpha, \Lambda_\varphi + \delta) t} - e^{-\pi i (\alpha, \Lambda_\varphi + \delta) t} \right).$$

Hence

$$\dim \varphi = \lim_{t \to 0} \chi_\varphi(\operatorname{Exp} t\delta)$$

$$= \prod_{\alpha \in \Delta^+} \lim_{t \to 0} \frac{2i \cdot \sin \pi (\alpha, \Lambda_\varphi + \delta) t}{2i \cdot \sin \pi (\alpha, \delta) t}$$

$$= \prod_{\alpha \in \Delta^+} \frac{(\alpha, \Lambda_\varphi + \delta)}{(\alpha, \delta)}. \qquad \square$$

Remarks (i) Theorem 4 is a far-reaching generalization of Theorem 1.7.

(ii) Theorem 4 only settles the uniqueness aspect of the classification problem of irreducible representations.

(iii) The existence aspect amounts to determine which vector in Γ_0 can be realized as the highest weight of a complex irreducible representations of G. This depends on the structure of G and hence can only be satisfactorily answered after more structural theory of compact connected Lie groups, (cf. Lecture 5).

Lecture 5

Structural Theory of Compact Lie Algebras

A Lie algebra $\mathfrak{G}$ over $\mathbb{R}$ is called a *compact* Lie algebra if it can be realized as the Lie algebra of a compact Lie group G. Let us analyze the algebraic implications of the above rather geometric definition in order to obtain algebraic characterization of compact Lie algebras.

1. Characterization of Compact Lie Algebras

Lemma 1. *If $\mathfrak{G}$ is a compact Lie algebra, then there exists an inner product $(\ ,)$ on $\mathfrak{G}$ such that*

$$([X,Y],Z) + (Y,[X,Z]) \equiv 0, \tag{1}$$

for all X, Y, Z in $\mathfrak{G}$.

Proof: Suppose $\mathfrak{G}$ is the Lie algebra of a compact Lie group G. Then there exists an Ad_G-invariant inner product $(\ ,)$ on $\mathfrak{G}$. Let X, Y, Z be arbitrary elements of $\mathfrak{G}$. Then

$$(\mathrm{Ad}(\mathrm{Exp}\,tX)Y, \mathrm{Ad}(\mathrm{Exp}\,tX)Z) = (Y,Z), \quad t \in \mathbb{R}. \tag{2}$$

Differentiate the above equation at $t = 0$, one gets

$$([X, Y], Z) + (Y, [X, Z]) \equiv 0. \qquad \square$$

Definition An inner product (,) on $\mathfrak{G}$ is called *invariant* if it satisfies the above identity.

Theorem 1. *A compact Lie algebra $\mathfrak{G}$ splits, uniquely, into the direct sum of its center and its simple ideals, namely*

$$\mathfrak{G} = \mathfrak{C} \oplus \mathfrak{G}_1 \oplus \cdots \oplus \mathfrak{G}_l,$$

when $\mathfrak{G}_j$ are distinct simple ideals of $\mathfrak{G}$. Moreover, each component is itself a compact Lie algebra.

Proof: Let G be a compact Lie group with $\mathfrak{G}$ as its Lie algebra and assume that $\mathfrak{G}$ is already equipped with an invariant inner product. Set

$$\mathfrak{C} = \{X \in \mathfrak{G}; [X, \mathfrak{G}] = 0\} \quad \text{(the center of } \mathfrak{G}\text{)}$$
$$\text{and } \mathfrak{G}' = \mathfrak{C}^{\perp}. \tag{3}$$

Then, it follows easily from (1) and (3) that

$$[\mathfrak{G}, \mathfrak{G}'] \subset \mathfrak{G}'$$
$$[\mathfrak{G}, \mathfrak{G}]^{\perp} \subset \mathfrak{C}, \tag{4}$$

and hence $\mathfrak{G} = \mathfrak{C} \oplus \mathfrak{G}'$ as Lie algebra and moreover,

$$[\mathfrak{G}, \mathfrak{G}] = [\mathfrak{G}', \mathfrak{G}'] = \mathfrak{G}'. \tag{5}$$

Suppose $\mathfrak{G}_1$ is a simple ideal of $\mathfrak{G}'$. Then it is also a simple ideal of $\mathfrak{G}$ and it follows from (1) that

$$\mathfrak{G}'' = (\mathfrak{C} \oplus \mathfrak{G}_1)^{\perp}$$

is also an ideal of $\mathfrak{G}$, namely,

$$\mathfrak{G} = \mathfrak{C} \oplus \mathfrak{G}_1 \oplus \mathfrak{G}'' \quad \text{(as Lie algebra)}.$$

Let G_1, G'' be the connected Lie subgroups of G with $\mathfrak{G}_1$, $\mathfrak{G}''$ as their Lie algebras. Let $Z_G^o(G_1)$, $Z_G^o(G'')$ and Z^o be the connected centralizer of G_1, G'' and G respectively. Then it is easy to see that their Lie algebras are respectively

$$\mathfrak{C} \oplus \mathfrak{G}'', \quad \mathfrak{C} \oplus \mathfrak{G}_1 \quad \text{and} \quad \mathfrak{C},$$

and hence, $\mathfrak{G}_1$ and $\mathfrak{G}''$ are the respective Lie algebras of

$$Z_G^o(G'')/Z^o \quad \text{and} \quad Z_G^o(G_1)/Z^o,$$

which are clearly compact Lie groups. This proves that both $\mathfrak{G}_1$ and $\mathfrak{G}''$ are themselves compact Lie algebras and it is then easy to complete the proof by induction on the dimension of $\mathfrak{G}$.   □

Definition   The Cartan–Killing form of a Lie algebra $\mathfrak{G}$ is defined to be

$$B(X,Y) = \operatorname{tr} \operatorname{ad}_X \circ \operatorname{ad}_Y, \quad X,Y \in \mathfrak{G}. \tag{6}$$

Lemma 2. $B(X,Y)$ *is a symmetric bilinear form and*

$$B(AX, AY) = B(X,Y),$$

for any automorphism A of $\mathfrak{G}$.

Proof: It is straightforward to check that B is both symmetric and bilinear. Let A be an automorphism of $\mathfrak{G}$. Then $A[X,Y] = [AX, AY]$ simply means

$$A \cdot \operatorname{ad}_X = \operatorname{ad}_{AX} \cdot A \quad \text{or} \quad \operatorname{ad}_{AX} = A \cdot \operatorname{ad}_X \cdot A^{-1}.$$

Therefore

$$\begin{aligned}
B(AX, AY) &= \operatorname{tr} \operatorname{ad}_{AX} \cdot \operatorname{ad}_{AY} \\
&= \operatorname{tr} A \operatorname{ad}_X A^{-1} \cdot A \operatorname{ad}_Y A^{-1} \\
&= \operatorname{tr} \operatorname{ad}_X \cdot \operatorname{ad}_Y = B(X,Y). \quad \square
\end{aligned}$$

Corollary $B([X,Y], Z) + B(Y, [X,Z]) \equiv 0.$

Proof: $\operatorname{Exp}(t \operatorname{ad}_X)$ is a one-dimensional subgroup of automorphism of $\mathfrak{G}$. Hence

$$B(\operatorname{Exp}(t \operatorname{ad}_X) \cdot Y, \ \operatorname{Exp}(t \operatorname{ad}_X) \cdot Z) \equiv B(Y, Z), \quad t \in \mathbb{R}.$$

Differentiate the above equation at $t = 0$, one gets

$$B([X,Y], Z) + B(Y, [X,Z]) \equiv 0. \quad \square$$

Lemma 3.   *If $\mathfrak{G}$ is a simple compact Lie algebra, then B is negative definite.*

Proof: Equipped $\mathfrak{G}$ with an invariant inner product. Then ad_X is an anti-symmetric linear transformation of $\mathfrak{G}$ and hence all its eigenvalues are purely imaginary. Therefore

$$B(X,X) = \operatorname{tr}(\operatorname{ad}_X)^2 \leq 0,$$

and equals to zero only when $\operatorname{ad}_X = 0$, i.e. $X = 0$, namely, B is negative definite.   □

Lemma 4.   *If $\mathfrak{G}$ is a Lie algebra with negative definite Cartan–Killing form and D is a derivation of $\mathfrak{G}$, then there exists $Z \in \mathfrak{G}$ with $D = \mathrm{ad}_Z$.*

Proof: Recall that $\mathrm{Exp}(tD)$ is a one-parameter subgroup of automorphisms of $\mathfrak{G}$ if and only if D is a derivation of $\mathfrak{G}$, namely,

$$D[X,Y] = [DX,Y] + [X,DY], \quad X,Y \in \mathfrak{G}. \tag{7}$$

Therefore, the set of all derivations of $\mathfrak{G}$, says $\mathfrak{D}$, is a Lie subalgebra of $\mathfrak{gl}(\mathfrak{G})$ and it contains $\mathrm{ad}\mathfrak{G}$ as one of its Lie subalgebras. Moreover, for $D \in \mathfrak{D}$ and $X \in \mathfrak{G}$,

$$\begin{aligned}
[D, \mathrm{ad}_X]Y &= D \cdot \mathrm{ad}_X(Y) - \mathrm{ad}_X(DY) \\
&= D[X,Y] - [X,DY] = [DX,Y] = \mathrm{ad}_{DX}(Y),
\end{aligned} \tag{8}$$

namely,

$$[D, \mathrm{ad}_X] = \mathrm{ad}_{DX}, \quad [\mathfrak{D}, \mathrm{ad}\mathfrak{G}] \subset \mathrm{ad}\mathfrak{G}.$$

Let $\tilde{B}$ and B be the Cartan–Killing form of $\mathfrak{D}$ and $\mathrm{ad}\mathfrak{G}$ respectively. Then it follows from $[\mathfrak{D}, \mathrm{ad}\mathfrak{G}] \subset \mathrm{ad}\mathfrak{G}$ that

$$\tilde{B}(X,Y) = B(X,Y) \quad \text{for } X,Y \in \mathrm{ad}\mathfrak{G}.$$

Set

$$I = \{D \in \mathfrak{D}; \tilde{B}(D, \mathrm{ad}\mathfrak{G}) = 0\}.$$

Then it is easy to see that I is also an ideal of $\mathfrak{D}$ and it follows from the negative definiteness of the Cartan–Killing form of $\mathfrak{G}$ that $\mathrm{ad} : \mathfrak{G} \to \mathrm{ad}\mathfrak{G}$ is an isomorphism and $I \cap \mathrm{ad}\mathfrak{G} = \{0\}$, $[I, \mathrm{ad}\mathfrak{G}] = \{0\}$.

Let D be an arbitrary element of I. Then

$$\begin{aligned}
[D, \mathrm{ad}_X] &= \mathrm{ad}_{DX} = 0 \quad \text{for all } X \in \mathfrak{G} \\
&\Rightarrow DX = 0 \quad \text{for all } X \in \mathfrak{G} \Rightarrow I = \{0\} \\
&\Rightarrow \mathfrak{D} = \mathrm{ad}\mathfrak{G}. \qquad \square
\end{aligned}$$

Theorem 2.   *A simple Lie algebra $\mathfrak{G}$ is compact if and only if either* (i) *its Cartan–Killing form is negative definite, or* (ii) *it has an invariant inner product.*

Proof: It is quite obvious that (i) ⇔ (ii). The "only if" part is already proved in Lemma 3. Therefore, what remains to be proved is that (i) implies that $\mathfrak{G}$ is compact.

Let $\mathrm{ad}\mathfrak{G}$ be the image of $\mathrm{ad} : \mathfrak{G} \to \mathfrak{gl}(\mathfrak{G})$. The simplicity of $\mathfrak{G}$ implies that $\mathrm{ad}\mathfrak{G} \cong \mathfrak{G}$ and condition (i) or (ii) implies that $\mathrm{ad}\mathfrak{G}$ is a Lie subalgebra of the Lie algebra of anti-symmetric matrices, namely, the Lie algebra of the orthogonal group of $\mathfrak{G}$, $O(\mathfrak{G})$. On the other hand, $\mathrm{ad}\mathfrak{G}$ is also the Lie algebra of all derivations of $\mathfrak{G}$ and hence, it is exactly the Lie algebra of the automorphisms groups of $\mathfrak{G}$, $\mathrm{Aut}(\mathfrak{G})$, which is clearly a *closed subgroup* of $O(\mathfrak{G})$. Therefore, $\mathfrak{G} \cong \mathrm{ad}\mathfrak{G}$ is the Lie algebra of the compact Lie group $\mathrm{Aut}(\mathfrak{G})$.□

Theorem 3 (H. Weyl). *Let G be a compact connected Lie group and $\tilde{G}$ be its universal covering group. If its Lie algebra $\mathfrak{G}$ has no center, then $\tilde{G}$ is also compact.*

Remarks   (i) In the special case of rank one, it follows from the classification theorem that $\tilde{G} = S^3$. Hence the above theorem is a generalization of the above known special case.

(ii) In case $\mathfrak{G}$ has non-trivial center, namely, $\mathfrak{G} = \mathfrak{C} \oplus \mathfrak{G}'$, $\dim \mathfrak{C} = d > 0$, then $\tilde{G}$ contains a factor of $\mathbb{R}^d$ and hence non-compact. Therefore, the above theorem, in fact, asserts that $\tilde{G}$ is compact if and only if $\mathfrak{G}$ has no center.

Proof of Theorem 3:   Let $\pi : \tilde{G} \to G$ be the universal covering of G. We shall first equip G with a bi-invariant Riemannian metric and $\tilde{G}$ with the induced covering metric which is, of course, also bi-invariant. Let T be a maximal torus of G and $G(t_0)$, $t_0 \in U \cap T$, be an orbit of the generic type which is contained in an *evenly covered neighborhood* of the identity in G (cf. Section 1 of Lecture 3). Let $\mathfrak{h}$ be the Lie algebra of T, $\tilde{T} = \mathrm{Exp}\,\mathfrak{h}$ in $\tilde{G}$ and $\tilde{t}_0$ be the unique lifting of t_0 in the neighborhood of identity in $\tilde{G}$. Then, it is easy to see that $\tilde{G}(\tilde{t}_0)$ is the unique lifting of $G(t_0)$ in the neighborhood of identity in $\tilde{G}$. The following commutative diagram summarizes the above situation:

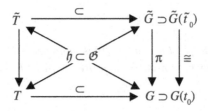

where both $\{T, G(t_0)\}$ and $\{\tilde{T}, \tilde{G}(\hat{t}_0)\}$ intersect transversally and perpendicularly at t_0 and $\hat{t}_0$, respectively.

Since $\tilde{T}$ is again *totally geodesic* in $\tilde{G}$ and $\tilde{G}$ is complete, it follows from the same simple geometric reason that $\tilde{T}$ intersects every conjugacy class of $\tilde{G}$. In particular,

$$\tilde{T} \supset Z(\tilde{G}) \supset \ker \pi,$$

namely, $\tilde{T} = \pi^{-1}(T)$ and hence $\tilde{G}$ is a *finite* sheet covering if and only if $\tilde{T}$ is still a torus, i.e., still *compact*. Suppose the contrary that $\tilde{T} \simeq T_1 \times \mathbb{R}^d$, $d > 0$. Then, for each pair of root $\pm\alpha \in \Delta$,

$$\tilde{G}_\alpha = \pi^{-1}(G_\alpha) \sim S^3 \times \tilde{T}_\alpha$$
$$\Rightarrow \mathbb{R}^d \subset \tilde{T}_\alpha = \mathrm{Exp}\,\mathfrak{h}_\alpha,$$

where $\mathfrak{h}_\alpha$ is the kernel of $\alpha : T \to \mathbb{R}$. Therefore, the Lie algebra of $\mathbb{R}^d$ lies in

$$\bigcap\{\mathfrak{h}_\alpha; \alpha \in \Delta\} = \text{the center of } \mathfrak{G},$$

which is clearly a contradiction to the assumption that $\mathfrak{G}$ has no center. □

Summarizing the above discussion on compact Lie algebras, we state the result of this section as follows:

(1) A Lie algebra (over $\mathbb{R}$) is compact if and only if it possesses an *invariant* inner product.
(2) Every compact Lie algebra $\mathfrak{G}$ can be uniquely decomposed into the direct sum of its center and a semi-simple compact Lie algebra, namely, $\mathfrak{G} = \mathfrak{C} \oplus \mathfrak{G}'$, where $\mathfrak{G}'$ is a direct sum of simple compact Lie algebras.
(3) A center-less Lie algebra $\mathfrak{G}$ is compact if and only if its Cartan–Killing form is negative definite.
(4) For every center-less compact Lie algebra $\mathfrak{G}$, its connected automorphism group is a compact linear group with ad$\mathfrak{G}$ as its Lie algebra; the simply connected Lie group with $\mathfrak{G}$ as its Lie algebra is also a compact Lie group.

2. Cartan Decomposition and Structural Constants of Compact Lie Algebras

By Theorem 1, the structure of a compact Lie algebra $\mathfrak{G}$ can easily be reduced to that of its simple components. Hence, for simplicity and without loss of generality, we shall always assume that a compact Lie algebra $\mathfrak{G}$ is simple to begin with in the following discussion.

Let $\mathfrak{G}$ be a given compact simple (or semi-simple) Lie algebra and G be either the connected automorphism group of $\mathfrak{G}$ (with ad$\mathfrak{G} \simeq \mathfrak{G}$ as its Lie algebra) or the simply connected Lie group with $\mathfrak{G}$ as its Lie algebra (by Theorem 3, it is compact). Recall that the Cartan–Killing form of $\mathfrak{G}$ is negative definite and hence it provides an *intrinsic* inner product on $\mathfrak{G}$, namely, $(X, Y) = -B(X, Y) = -\operatorname{tr} \operatorname{ad}_X \cdot \operatorname{ad}_Y$. Let T be a maximal torus of G, $\mathfrak{h}$ be its Lie algebra (i.e. a Cartan subalgebra of $\mathfrak{G}$) and $\Delta \subset \mathfrak{h}^*$ be the root system of $\mathfrak{G}$. In fact, it is slightly more convenient to use the above intrinsic inner product to identify $\mathfrak{h}^*$ with $\mathfrak{h}$ and to consider Δ as a subset of $\mathfrak{h}$ itself.

Cartan Decomposition of $\mathfrak{G} \otimes \mathbb{C}$ and $\mathfrak{G}$ Recall that the adjoint transformation

$$\operatorname{Ad} : G \times G \to G$$

is actually the geometric representation of the *totality of the non-commutativity* of G. The adjoint representation of G

$$\operatorname{Ad} : G \times \mathfrak{G} \to \mathfrak{G}$$

and the adjoint representation of $\mathfrak{G}$

$$\operatorname{ad} : \mathfrak{G} \times \mathfrak{G} \to \mathfrak{G}$$

are exactly the two stages of linearization of the above adjoint transformation. The restriction of $\operatorname{Ad} \otimes \mathbb{C}$ to T gives the following decomposition of $\mathfrak{G} \otimes \mathbb{C}$, namely,

$$\mathfrak{G} \otimes \mathbb{C} = \mathfrak{h} \otimes \mathbb{C} \oplus \sum_{\alpha \in \Delta} \mathbb{C}_\alpha \qquad (9)$$

such that

$$\begin{cases} \operatorname{Ad}(\operatorname{Exp} tH) \cdot Z_\alpha = e^{2\pi i(\alpha, H)t} \cdot Z_\alpha, \\ [H, Z_\alpha] = \operatorname{ad}_H \cdot Z_\alpha = 2\pi i(\alpha, H) \cdot Z_\alpha, \end{cases} \qquad (10')$$

for all $H \in \mathfrak{h}$ and $Z_\alpha \in \mathbb{C}_\alpha$. Correspondingly, the restriction of Ad_G to T gives the following Cartan decomposition of $\mathfrak{G}$, namely,

$$\mathfrak{G} = \mathfrak{h} \oplus \sum_{\pm\alpha \in \Delta} \mathbb{R}^2_{(\pm\alpha)}, \qquad (10)$$

such that

$$\begin{cases} \operatorname{Ad}(\operatorname{Exp} tH) \cdot X_\alpha = \cos 2\pi(\alpha, H) t \cdot X_\alpha + \sin 2\pi(\alpha, H) t \cdot Y_\alpha, \\ \operatorname{Ad}(\operatorname{Exp} tH) \cdot Y_\alpha = -\sin 2\pi(\alpha, H) t \cdot X_\alpha + \cos 2\pi(\alpha, H) t \cdot Y_\alpha, \\ [H, X_\alpha] = \operatorname{ad}_H \cdot X_\alpha = 2\pi(\alpha, H) \cdot Y_\alpha, \\ [H, Y_\alpha] = \operatorname{ad}_H \cdot Y_\alpha = -2\pi(\alpha, H) \cdot X_\alpha, \end{cases} \quad (11')$$

for $H \in \mathfrak{h}$ and orthonormal basis $\{X_\alpha, Y_\alpha\}$ in $\mathbb{R}^2_{\pm\alpha}$.

Lemma 5. *Let G_α be the connected Lie subgroup of G with $\mathfrak{G}_\alpha = \mathfrak{h} \oplus \mathbb{R}^2_{\pm\alpha}$ as its Lie algebra (cf. Theorem 4.2). Then the restriction of $\operatorname{Ad}_G \otimes \mathbb{C}$ to G_α has the following decomposition into its complex irreducible components, namely*

$$\mathfrak{G} \otimes \mathbb{C} = \langle \alpha \rangle^\perp \oplus \{\langle \alpha \rangle \oplus \mathbb{C}_\alpha \oplus \mathbb{C}_{-\alpha}\}$$
$$\oplus \sum_{\alpha\text{-string}} \{\mathbb{C}_{\beta+p\alpha} \oplus \cdots \oplus \mathbb{C}_{\beta+q\alpha}\}, \quad (12)$$

where $\{\beta + j\alpha; q(\alpha, \beta) \leq j \leq p(\alpha, \beta)\}$ is the α-string in Δ passing through β.

Proof: $T \subset G_\alpha \subset G$, $\Delta(G_\alpha) = \{\pm\alpha\}$ and there is a covering homomorphism of $S^3 \times T_\alpha$ onto G_α. Therefore, an irreducible complex representation of G_α can also be considered as an irreducible complex representation of $S^3 \times T_\alpha$ via the pull-back, and hence, its weight system forms an α-*string* reflectionally symmetric with respect to the Lie algebra of T_α, i.e. $\mathfrak{h}_\alpha = \langle \alpha \rangle^\perp$, (cf. Theorem 4.2). Since the multiplicity of each root $\beta \in \Delta$ is 1, each root β belongs to a unique α-string of roots passing through it, namely,

$$\{\beta + j\alpha; q(\alpha, \beta) \leq j \leq p(\alpha, \beta)\}. \qquad \square$$

Remark In fact, the lengths of the above α-strings of roots are at most 3 and α-strings of length 3 only occur in the case $\Delta(G)$ is of G_2-type (cf. Lecture 6).

Lemma 6. *For a pair of roots $\alpha, \beta \in \Delta$, $\alpha + \beta \neq 0$,*

$$\begin{cases} [\mathbb{C}_\alpha, \mathbb{C}_\beta] = 0 & \text{if } \alpha + \beta \notin \Delta, \\ [\mathbb{C}_\alpha, \mathbb{C}_\beta] = \mathbb{C}_{\alpha+\beta} & \text{if } \alpha + \beta \in \Delta, \\ \dfrac{2(\beta, \alpha)}{(\alpha, \alpha)} = -(p(\alpha, \beta) + q(\alpha, \beta)). \end{cases} \quad (13')$$

Proof: Let Z_α, Z_β be non-zero elements of $\mathbb{C}_\alpha$, $\mathbb{C}_\beta$, respectively, and H be an arbitrary element of $\mathfrak{h}$. Then

$$[H, Z_\alpha] = 2\pi i(\alpha, H) \cdot Z_\alpha,$$
$$[H, Z_\beta] = 2\pi i(\beta, H) \cdot Z_\beta,$$

and hence

$$[H, [Z_\alpha, Z_\beta]] = [[H, Z_\alpha], Z_\beta] + [Z_\alpha, [H, Z_\beta]]$$
$$= 2\pi i(\alpha + \beta, H) \cdot [Z_\alpha, Z_\beta].$$

Therefore, $[Z_\alpha, Z_\beta] = 0$ if $\alpha + \beta \notin \Delta$ and $[Z_\alpha, Z_\beta] \in \mathbb{C}_{\alpha+\beta}$ if $\alpha + \beta \in \Delta$, and moreover $[Z_\alpha, Z_\beta] \neq 0$ in the latter case. For otherwise,

$$\mathbb{C}_{\beta+q\alpha} \oplus \cdots \oplus \mathbb{C}_\beta$$

already forms a G_α-invariant subspace of $\mathfrak{G} \otimes \mathbb{C}$, which contradicts Lemma 5.

Finally, since $\{\beta + j\alpha; q(\alpha, \beta) \leq j \leq p(\alpha, \beta)\}$ is an α-string of roots reflectionally symmetric with respect to $\langle \alpha \rangle^\perp$, one has

$$\beta + q(\alpha, \beta)\alpha = \beta + p(\alpha, \beta)\alpha - \frac{2(\beta + p(\alpha, \beta)\alpha, \alpha)}{(\alpha, \alpha)} \cdot \alpha,$$

namely,

$$\frac{2(\beta, \alpha)}{(\alpha, \alpha)} \cdot \alpha = -(p(\alpha, \beta) + q(\alpha, \beta)) \cdot \alpha.$$

This proves that

$$\frac{2(\beta, \alpha)}{(\alpha, \alpha)} = -(p(\alpha, \beta) + q(\alpha, \beta)). \qquad \square$$

Structural Constants Recall that $\mathfrak{G}$ is a given compact simple (or semi-simple) Lie algebra equipped with the intrinsic inner product $(X, Y) = -B(X, Y) = -\text{tr}\,\text{ad}_X \cdot \text{ad}_Y$, and

$$\mathfrak{G} \otimes \mathbb{C} = \mathfrak{h} \otimes \mathbb{C} \oplus \sum_{\alpha \in \Delta} \mathbb{C}_\alpha,$$

$$\mathfrak{G} = \mathfrak{h} \oplus \sum_{\pm\alpha \in \Delta} \mathbb{R}^2_{\pm\alpha}$$

are the Cartan decomposition of $\mathfrak{G} \otimes \mathbb{C}$ and $\mathfrak{G}$ respectively. Let Z_α be a unit vector of $\mathbb{C}_\alpha$ and $\{X_\alpha, Y_\alpha\}$ be an orthonormal basis of $\mathbb{R}^2_{(\pm\alpha)}$ such that

$Z_{-\alpha} = \bar{Z}_\alpha$ and

$$\begin{cases} X_\alpha = \dfrac{1}{\sqrt{2}}(Z_\alpha + Z_{-\alpha}), \\[2mm] Y_\alpha = \dfrac{i}{\sqrt{2}}(Z_\alpha - Z_{-\alpha}), \end{cases} \tag{14}$$

$$\begin{cases} Z_\alpha = \dfrac{1}{\sqrt{2}}(X_\alpha - iY_\alpha), \\[2mm] Z_{-\alpha} = \dfrac{1}{\sqrt{2}}(X_\alpha + iY_\alpha). \end{cases} \tag{14$'$}$$

Then, one has $[X_\alpha, Y_\alpha] \in \mathfrak{h}$ and

$$([X_\alpha, Y_\alpha], H) = (Y_\alpha, [H, X_\alpha]) = 2\pi(\alpha, H), \tag{15}$$

for all $H \in \mathfrak{h}$. Hence

$$\begin{cases} [X_\alpha, Y_\alpha] = 2\pi\alpha, \\[2mm] [Z_\alpha, Z_{-\alpha}] = 2\pi i\alpha. \end{cases} \tag{15$'$}$$

Lemma 7. *For $\alpha, \beta, \alpha + \beta \in \Delta$, set $N_{\alpha,\beta}$ to be the structural constant such that $[Z_\alpha, Z_\beta] = N_{\alpha,\beta} Z_{\alpha+\beta}$. Then*

(i) $N_{\alpha,\beta} = -N_{\beta,\alpha}$,

(ii) $N_{-\alpha,-\beta} = \bar{N}_{\alpha,\beta}$,

(iii) *If $\alpha + \beta + \gamma = 0$, then $N_{\alpha,\beta} = N_{\beta,\gamma} = N_{\gamma,\alpha}$.*

(iv) *If $\alpha + \beta + \gamma + \delta = 0$ and there are no opposite roots in the above four roots, then*

$$N_{\alpha,\beta}N_{\gamma,\delta} + N_{\beta,\gamma}N_{\alpha,\delta} + N_{\gamma,\alpha}N_{\beta,\delta} = 0.$$

(v) $|N_{\alpha,\beta}|^2 = N_{\alpha,\beta} \cdot N_{-\alpha,-\beta} = 2\pi^2 p(\alpha,\beta)(1 - q(\alpha,\beta)) \cdot (\alpha,\alpha)$.

Proof:

(i) $[Z_\beta, Z_\alpha] = -[Z_\alpha, Z_\beta] \Rightarrow N_{\beta,\alpha} = -N_{\alpha,\beta}$.

(ii) $[Z_{-\alpha}, Z_{-\beta}] = [\bar{Z}_\alpha, \bar{Z}_\beta] = \bar{N}_{\alpha,\beta} \cdot \bar{Z}_{\alpha+\beta} = \bar{N}_{\alpha,\beta} \cdot Z_{-(\alpha+\beta)}$
$\Rightarrow N_{-\alpha,-\beta} = \bar{N}_{\alpha,\beta}$.

(iii) Suppose that $\alpha, \beta, \gamma \in \Delta$ and $\alpha + \beta + \gamma = 0$. Then

$$N_{\alpha,\beta} = (N_{\alpha,\beta} \cdot Z_{-\gamma}, Z_\gamma) = ([Z_\alpha, Z_\beta], Z_\gamma)$$
$$= (Z_\alpha, [Z_\beta, Z_\gamma]) = (Z_\alpha, N_{\beta,\gamma}Z_{-\alpha}) = N_{\beta,\gamma}.$$

Hence $N_{\alpha,\beta} = N_{\beta,\gamma} = N_{\gamma,\alpha}$.

(iv) Let α, β, γ, $\delta \in \Delta$, $\alpha + \beta + \gamma + \delta = 0$ and there are no opposite pairs among them. Suppose that $\beta + \gamma \in \Delta$. Then $\alpha + (\beta + \gamma) + \delta = 0$ and hence

$$[Z_\alpha, [Z_\beta, Z_\gamma]] = N_{\beta,\gamma} N_{\alpha,\beta+\gamma} Z_{-\delta} = -N_{\beta,\gamma} N_{\alpha,\delta} Z_{-\delta}.$$

[The above still holds if we set $N_{\beta,\gamma} = 0$ for the case $\beta + \gamma \notin \Delta$.] Therefore, it follows from the Jacobi identity that

$$-\{N_{\beta,\gamma} N_{\alpha,\delta} + N_{\gamma,\alpha} N_{\beta,\delta} + N_{\alpha,\beta} N_{\gamma,\delta}\} \cdot Z_{-\delta}$$
$$= [Z_\alpha, [Z_\beta, Z_\gamma]] + [Z_\beta, [Z_\gamma, Z_\alpha]] + [Z_\gamma, [Z_\alpha, Z_\beta]] = 0,$$

which clearly implies that

$$N_{\alpha,\beta} N_{\gamma,\delta} + N_{\beta,\gamma} N_{\alpha,\delta} + N_{\gamma,\alpha} N_{\beta,\delta} = 0.$$

(v)

$$[Z_{-\alpha}, [Z_\alpha, Z_{\beta+q\alpha}]] = [[Z_{-\alpha}, Z_\alpha], Z_{\beta+q\alpha}] + 0$$
$$= [-2\pi i \alpha, Z_{\beta+q\alpha}] = -(2\pi i)^2 \cdot (\alpha, \beta + q\alpha) \cdot Z_{\beta+q\alpha}$$
$$= 4\pi^2 \cdot \frac{1}{2}(q - p) \cdot (\alpha, \alpha) \cdot Z_{\beta+q\alpha}.$$

[Notice that $[Z_{-\alpha}, Z_{\beta+q\alpha}] = 0$ and $(\alpha, \beta) = -(p + q)(\alpha, \alpha)/2$.] Set $Z_q = Z_{\beta+q\alpha}$ and inductively $Z_{j+1} = [Z_\alpha, Z_j]$. Then

$$[Z_{-\alpha}, [Z_\alpha, Z_{j+1}]] = [[Z_{-\alpha}, Z_\alpha], Z_{j+1}] + [Z_\alpha, [Z_{-\alpha}, Z_{j+1}]]$$
$$= [-2\pi i \alpha, Z_{j+1}] + [Z_\alpha, [Z_{-\alpha}, [Z_\alpha, Z_j]]].$$

Therefore, it is quite straightforward to prove by induction that

$$[Z_{-\alpha}, [Z_\alpha, Z_j]] = 2\pi^2 (j - p)(1 - q + j)(\alpha, \alpha) \cdot Z_j.$$

In particular, one has

$$[Z_{-\alpha}, [Z_\alpha, Z_\beta]] = 2\pi^2 p(q - 1)(\alpha, \alpha) Z_\beta,$$

and hence

$$2\pi^2 p(q - 1)(\alpha, \alpha) = N_{\alpha,\beta} N_{-\alpha,\alpha+\beta}$$
$$= N_{\alpha,\beta} N_{-\beta,-\alpha} = -|N_{\alpha,\beta}|^2,$$

namely

$$|N_{\alpha,\beta}|^2 = 2\pi^2 p(1 - q)(\alpha, \alpha). \qquad \Box$$

Theorem 4 (Chevalley). *Let $\mathfrak{G}$ be a simple compact Lie algebra, Δ be its root system and $\pi = \{\alpha_1, \ldots, \alpha_k\}$ be a chosen system of simple roots. Then, there exists a basis of the Cartan decomposition of $\mathfrak{G} \otimes \mathbb{C}$*

$$\{H_j \in \mathfrak{h}, 1 \leq j \leq k; Z'_\alpha \in \mathbb{C}_\alpha, \alpha \in \Delta\},$$

with the following structural constants:

(i) $[H_j, Z'_\alpha] = \frac{2(\alpha_j, \alpha)}{(\alpha_j, \alpha_j)} i Z'_\alpha$,

(ii) $[Z'_\alpha, Z'_{-\alpha}] = iH_\alpha = \frac{i\alpha}{\pi(\alpha,\alpha)}$, H_α *is an integral linear combination of H_j,* $1 \leq j \leq k$,

(iii) $[Z'_\alpha, Z'_\beta] = 0$ *if $\alpha + \beta \notin \Delta$,*

(iv) $[Z'_\alpha, Z'_\beta] = \pm(1-q)Z_{\alpha+\beta}$ *if $\alpha, \beta, \alpha+\beta \in \Delta$ and $\{\beta + j\alpha; q \leq j \leq p\}$ is the α-string in Δ containing β.*

Proof: (i) Set

$$H_\alpha = \frac{\alpha}{\pi(\alpha,\alpha)} \quad \text{and} \quad H_j = H_{\alpha_j}, 1 \leq j \leq k,$$

$$Z'_\alpha = \frac{1}{\pi\sqrt{2(\alpha,\alpha)}}Z_\alpha, \quad X'_\alpha = \frac{1}{\pi\sqrt{2(\alpha,\alpha)}}X_\alpha, \quad Y'_\alpha = \frac{1}{\pi\sqrt{2(\alpha,\alpha)}}Y_\alpha.$$

Then

$$[X'_\alpha, Y'_\alpha] = \frac{2\pi\alpha}{2\pi^2(\alpha,\alpha)} = \frac{\alpha}{\pi(\alpha,\alpha)} = H_\alpha,$$

$$[Z'_\alpha, Z'_{-\alpha}] = \frac{2\pi i\alpha}{2\pi^2(\alpha,\alpha)} = \frac{\alpha i}{\pi(\alpha,\alpha)} = iH_\alpha,$$

$$[H_\alpha, X'_\beta] = 2\pi(\beta, H_\alpha)Y'_\beta = \frac{2(\alpha,\beta)}{(\alpha,\alpha)}Y'_\beta,$$

$$[H_\alpha, Y'_\beta] = -2\pi(\beta, H_\alpha)X'_\beta = -\frac{2(\alpha,\beta)}{(\alpha,\alpha)}X'_\beta,$$

$$[H_\alpha, Z'_\beta] = 2\pi i(\beta, H_\alpha)Z'_\beta = \frac{2(\alpha,\beta)}{(\alpha,\alpha)}iZ'_\beta.$$

(ii) Set $N'_{\alpha,\beta} = 0$ if $\alpha + \beta \notin \Delta$, and

$$[Z'_\alpha, Z'_\beta] = N'_{\alpha,\beta}Z'_{\alpha+\beta},$$

if α, β, $\alpha + \beta \in \Delta$. Then

$$N'_{\alpha,\beta} = \frac{|\alpha + \beta|}{\sqrt{2\pi}|\alpha||\beta|}N_{\alpha,\beta},$$

and hence

$$|N'_{\alpha,\beta}|^2 = \frac{|\alpha + \beta|^2}{2\pi^2 |\alpha|^2 |\beta|^2} |N_{\alpha,\beta}|^2$$

$$= p(1 - q)\frac{|\alpha + \beta|^2}{|\beta|^2} \qquad \text{[by (v) of Lemma 7]}.$$

On the other hand, it is easy to check the list of root systems of rank 2 that

$$p\frac{|\alpha + \beta|^2}{|\beta|^2} = (1 - q)$$

holds in general. Therefore

$$|N'_{\alpha,\beta}|^2 = (1 - q)^2.$$

(iii) For $1 \le i, j \le k$, one has

$$r_i(H_j) = H_j - \frac{2(\alpha_i, H_j)}{(\alpha_i, \alpha_i)}\alpha_i$$

$$= H_j - \frac{2(\alpha_i, \alpha_j)}{\pi(\alpha_i, \alpha_i)(\alpha_j, \alpha_j)}\alpha_i = H_j - \frac{2(\alpha_i, \alpha_j)}{(\alpha_j, \alpha_j)}H_i.$$

Since every H_α, $\alpha \in \Delta$ can be transformed into an H_j by a suitable sequence of such reflections, it is easy to see that H_α is always an integral linear combination of $\{H_j; 1 \le j \le k\}$.

(iv) Observe that one may still adjust each pair $\{Z'_\alpha, Z'_{-\alpha}\}$ by a factor of unit complex numbers without disturbing all the above result, namely

$$\{Z'_\alpha, Z'_{-\alpha}\} \to \{e^{i\theta} Z'_\alpha, e^{-i\theta} Z'_{-\alpha}\}.$$

Therefore, the final part of the proof is to show that it is always possible to adjust all the pair consistently so that all $N'_{\alpha,\beta}$ are real! This can be accomplished by a simple procedure of inductive tune-up and (iv) of Lemma 7 as follows.

Let us again adopt an ordering in $\mathfrak{h}$ and set

$$\Delta_\gamma = \{\alpha \in \Delta; -\gamma < \alpha < \gamma\}.$$

Inductively, we shall assume that $\{Z'_\alpha, Z'_{-\alpha}\}$, $\alpha \in \Delta_\gamma$, have already been chosen such that

$$N'_{\alpha,\beta} \in \mathbb{R} \quad \text{for all} \quad \alpha, \beta, \alpha + \beta \in \Delta_\gamma.$$

If γ is an indecomposable positive root, then any choice of $\{Z'_\gamma, Z'_{-\gamma}\}$ will also satisfy

$$N'_{\alpha,\beta} \in \mathbb{R} \quad \text{for all} \quad \alpha, \beta, \alpha + \beta \in \Delta_\gamma \cup \{\pm\gamma\}.$$

Otherwise, let $\gamma = \alpha + \beta$ be the decomposition of γ with the smallest possible α. We shall re-adjust $\{Z'_\gamma, Z'_{-\gamma}\}$ so that

$$N'_{\alpha,\beta} \in \mathbb{R}^+.$$

Therefore, what remains to verify is that such an adjustment will make all other

$$N'_{\alpha_1,\beta_1}, \quad \alpha_1 + \beta_1 = \gamma$$

also real. Suppose $\gamma = \alpha_1 + \beta_1$ is another decomposition of γ. Then $\alpha + \beta + (-\alpha_1) + (-\beta_1) = 0$ and Lemma 7 (iv) applies. Hence

$$N_{\alpha,\beta} N_{-\alpha_1,-\beta_1} + N_{\beta,-\alpha_1} N_{\alpha,-\beta_1} + N_{-\alpha,\alpha} N_{\beta,-\beta_1} = 0.$$

The other five structural constants are real, this certainly implies that $N_{-\alpha,-\beta_1} = \bar{N}_{\alpha,\beta}$ is also real. $\qquad\square$

Remark If one set $H'_j = i \cdot H_j$, then the structural constants of $\mathfrak{G} \otimes \mathbb{C}$ with respect to the basis $\{H'_j, 1 \leq j \leq k; Z'_\alpha, Z'_{-\alpha}, \alpha \in \Delta^+\}$ are all integers. This basis enable us to obtain a Lie algebra over a field of characteristic p. The above theorem is usually called the Chevalley's basis theorem.

Lecture 6

Classification Theory
of Compact Lie Algebras and
Compact Connected Lie Groups

Let G be a given compact connected Lie group and $\mathfrak{G}$ be its Lie algebra. Then, by Theorem 5.1, $\mathfrak{G}$ splits into the direct sum of its center and its simple ideals, namely

$$\mathfrak{G} = \mathfrak{C} \oplus \mathfrak{G}_1 \oplus \cdots \oplus \mathfrak{G}_l,$$

where each $\mathfrak{G}_j$ is also a compact Lie algebra. The connected Lie subgroup, G_j, with $\mathfrak{G}_j$ as its Lie algebra is a compact subgroup of G. Let Z^o be the connected center of G. Then it follows from the above direct sum decomposition of $\mathfrak{G}$ that

$$Z^o \times G_1 \times \cdots \times G_l \to G, \quad (g_0, g_1, \ldots, g_l) \mapsto g_0 \cdot g_1 \cdots g_l$$

is a covering homomorphism with a finite kernel. This enable us to reduce the classification of compact connected Lie group to that of simple compact Lie algebras and that of simply connected, simple compact Lie groups (cf. Theorem 5.3) together with the determination of their centers.

1. Classification of Simple Compact Lie Algebras

Let us first summarize the results on the structures of simple compact Lie algebras that have already been established in the previous lectures.

(1) The Cartan–Killing form, $B(X,Y) = \mathrm{trad}_X \cdot \mathrm{ad}_Y$, of a simple compact Lie algebra $\mathfrak{G}$ is negative definite. Hence $\mathfrak{G}$ has an *intrinsic inner product*, $(X,Y) = -B(X,Y)$, which is *invariant*, i.e. $([X,Y],Z) + (Y,[X,Z]) \equiv 0$.

(2) It follows from the maximal tori theorem that any two maximal Abelian subalgebras of a given compact Lie algebra $\mathfrak{G}$ are conjugate under the action of Ad_G. Therefore, the *geometric properties of the root system*, Δ, are independent of the choice of the Cartan subalgebra $\mathfrak{h}$ (or the maximal torus T) and hence are, in fact, *structure invariants* of $\mathfrak{G}$.

(3) The Weyl group W acts on the Cartan subalgebra $\mathfrak{h}$ as a group generated by the reflections $\{r_\alpha; \pm\alpha \in \Delta\}$ where $r_\alpha(H) = H - \frac{2(\alpha,H)}{(\alpha,\alpha)}\alpha$, $H \in \mathfrak{h}$. It acts simply transitively on the set of chambers. Therefore, any two simple root systems (based on the choices of different Weyl chambers) of a given root system are W-conjugate.

(4) Theorem 5.4 has already gone a long way in determining the structure of a simple compact Lie algebra, $\mathfrak{G}$, *solely* in terms of the *homothetic property* of Δ. The following classification theorem is actually a slight up-grading of Theorem 5.4.

Theorem 1. *Two simple compact Lie algebras $\mathfrak{G}$ and $\mathfrak{G}'$ are isomorphic if and only if their simple root systems π and π' are homothetic, namely,*

$$\mathfrak{G} \cong \mathfrak{G}' \Leftrightarrow \pi \sim \pi'.$$

Proof: Let $\iota : \mathfrak{G} \cong \mathfrak{G}'$ be a given isomorphism of $\mathfrak{G}$ onto $\mathfrak{G}'$, $\mathfrak{h}$ and $\mathfrak{h}'$ be given Cartan subalgebras of $\mathfrak{G}$ and $\mathfrak{G}'$ respectively. Then $\iota(\mathfrak{h})$ and $\mathfrak{h}'$ are two maximal Abelian subalgebras of $\mathfrak{G}'$ and hence there exists, by Corollary 3.1, an adjoint automorphism $\sigma : \mathfrak{G}' \to \mathfrak{G}'$ such that $\sigma\iota(\mathfrak{h}) = \mathfrak{h}'$. Therefore $\sigma\iota$ maps the root system, Δ, of $\mathfrak{G}$ with respect to $\mathfrak{h}$ *isometrically* onto the root system, Δ', of $\mathfrak{G}'$ with respect to $\mathfrak{h}'$. Let π be a chosen simple root system in Δ. Then $\sigma\iota(\pi)$ is also a simple root system in Δ' which is W'-conjugate to any other simple root system π' in Δ'. Hence, π and π' must be *isometric*.

Next let us proceed to prove that $\pi \sim \pi'$ implies $\mathfrak{G} \cong \mathfrak{G}'$. We shall denote the corresponding element of $\alpha_j \in \pi$ by α'_j, namely

$$\frac{2(\alpha_i, \alpha_j)}{(\alpha_i, \alpha_i)} = \frac{2(\alpha'_i, \alpha'_j)}{(\alpha'_i, \alpha'_i)}, \quad 1 \le i, j \le k.$$

It is straightforward to check that the above homothety, $\alpha_j \leftrightarrow \alpha'_j$, extends linearly to an equivariant isomorphism between $(W, \mathfrak{h})$ and $(W', \mathfrak{h}')$ whose restriction to Δ is, of course, also a homothety between Δ and Δ'. We shall choose the orderings on $\mathfrak{h}'$ to be compatible with that of $\mathfrak{h}$ and shall denote the corresponding root of $\alpha \in \Delta$ simply by $\alpha' \in \Delta'$.

Let $\{H_j, 1 \le j \le k; Z_\alpha, \alpha \in \Delta\}$ and $\{H'_j, 1 \le j \le k, Z'_\alpha, \alpha' \in \Delta'\}$ be respectively the Chevalley basis of $\mathfrak{G} \otimes \mathbb{C}$ and $\mathfrak{G}' \otimes \mathbb{C}$ such that

$$N_{\alpha, \beta} = (1 - q) = N'_{\alpha', \beta'},$$

whenever $\gamma = \alpha + \beta$ (resp. $\gamma' = \alpha' + \beta'$) is the decomposition of γ (resp. γ') with the *smallest* possible α (resp. α'). Then it follows from (iv) of Lemma 5.7 that

$$N_{\alpha, \beta} = N'_{\alpha', \beta'},$$

for all α, β, $\alpha + \beta \in \Delta$. Therefore, the homothety

$$\iota : \mathfrak{h} \to \mathfrak{h}', \quad \iota(H_j) = H'_j, \quad 1 \le j \le k$$

extends to an isomorphism of complex Lie algebras

$$\iota^* : \mathfrak{G} \otimes \mathbb{C} \to \mathfrak{G}' \otimes \mathbb{C}, \quad \iota^*(Z_\alpha) = Z'_{\alpha'}$$

and moreover, $\iota^*(\bar{Z}) = \overline{\iota^*(Z)}$ for all $Z \in \mathfrak{G} \otimes \mathbb{C}$.

Hence, the restriction of ι^* to $\mathfrak{G}$ is an isomorphism of $\mathfrak{G}$ onto $\mathfrak{G}'$. In fact, it maps the vectors

$$X_\alpha = \frac{1}{\sqrt{2}}(Z_\alpha + Z_{-\alpha}) \quad \text{and} \quad Y_\alpha = \frac{i}{\sqrt{2}}(Z_\alpha - Z_{-\alpha})$$

to

$$X'_{\alpha'} = \frac{1}{\sqrt{2}}(Z'_{\alpha'} + Z'_{-\alpha'}) \quad \text{and} \quad Y'_{\alpha'} = \frac{i}{\sqrt{2}}(Z'_{\alpha'} - Z'_{-\alpha'}),$$

respectively. $\qquad\qquad\qquad\qquad\qquad\qquad\qquad\qquad\qquad\qquad\square$

In the special case of $\mathfrak{G} = \mathfrak{G}'$, it is not difficult to refine the above isomorphism theorem into an automorphism theorem.

Theorem 2. *Let $\mathfrak{G}$ be a given simple compact Lie algebra, $\mathfrak{h}$ be a Cartan subalgebra of $\mathfrak{G}$ and π be a simple root system of $\mathfrak{G}$. Let $\mathrm{Aut}(\mathfrak{G})$ be the group of all automorphisms of $\mathfrak{G}$, $\mathrm{Ad}(\mathfrak{G})$ be the connected Lie subgroup of $Gl(\mathfrak{G})$ with $\mathrm{ad}(\mathfrak{G})$ as its Lie algebra. Then*

- (i) $\mathrm{Ad}(\mathfrak{G})$ *is exactly the connected component of the identity in* $\mathrm{Aut}(\mathfrak{G})$, *i.e.* $\mathrm{Ad}(\mathfrak{G}) = \mathrm{Aut}^o(\mathfrak{G})$.
- (ii) $\mathrm{Aut}(\mathfrak{G})/\mathrm{Ad}(\mathfrak{G}) \cong \mathrm{Isom}(\pi)$, *the group of isometries of* π.

Proof: (i) Let $\mathfrak{D}$ be the Lie algebra of $\mathrm{Aut}(\mathfrak{G})$ and D be an arbitrary element of $\mathfrak{D}$. Then

$$\mathrm{Exp}\, tD \cdot [X, Y] = [\mathrm{Exp}\, tD \cdot X, \mathrm{Exp}\, tD \cdot Y]; \quad X, Y \in \mathfrak{G}.$$

Therefore, by differentiation at $t = 0$,

$$D \cdot [X, Y] = [DX, Y] + [X, DY], \quad X, Y \in \mathfrak{G},$$

namely, D is a derivation of $\mathfrak{G}$. Hence, by Lemma 5.4, $\mathfrak{D} = \mathrm{ad}(\mathfrak{G})$ and hence $\mathrm{Aut}^o(\mathfrak{G}) = \mathrm{Ad}(\mathfrak{G})$.

(ii) For the proof of the second assertion, it is convenient to identify $\mathfrak{G}$ with $\mathrm{ad}(\mathfrak{G})$ and denote $\mathrm{Ad}(\mathfrak{G})$ simply by G. Let T be the maximal torus of G with the given $\mathfrak{h}$ as its Lie algebra and a be an arbitrary element of $\mathrm{Aut}(\mathfrak{G})$. Then $a(\mathfrak{h})$ is again a maximal Abelian subalgebra of $\mathfrak{G}$ and, by Corollary 3.1, there exists $g \in G$ such that $ga(\mathfrak{h}) = \mathfrak{h}$, $ga(\Delta) = \Delta$. Hence, ga permutes the set of chambers. Let C_0 be the Weyl chamber corresponding to the chosen simple root system π. Then, by Lemma 4.1, there exists an element $n \in N(T)$ such that $nga(C_0) = C_0$. Moreover, if $a \in G$, then $ga(\mathfrak{h}) = \mathfrak{h}$ implies that $ga \in N(T)$ and it follows from the simple transitivity of the W-action on the set of chambers that $nga(C_0) = C_0$ implies that $nga \in T$. Therefore, in the case $a \in G$, the restriction of the above nga to C_0 (resp. π) is the identity map.

The above discussion shows that $\mathrm{Aut}(\mathfrak{G})/\mathrm{Ad}(\mathfrak{G})$ has a natural induced isometric action on C_0 as well as on π, namely, it defines a homomorphism

$$\rho : \mathrm{Aut}(\mathfrak{G})/\mathrm{Ad}(\mathfrak{G}) \to \mathrm{Isom}(\pi).$$

The above homomorphism is surjective because any isometry of π can be extended to an automorphism of $\mathfrak{G}$, by Theorem 1.

Suppose that $a \in \mathrm{Aut}(\mathfrak{G})$ and its restriction to C_0 is the identity map. Then, in the Cartan decomposition of $\mathfrak{G} \otimes \mathbb{C}$, $\mathbb{C}_{\alpha_j}$, $1 \leq j \leq k$, are all invariant subspaces of a. Hence, there exists suitable $\{\theta_j, 1 \leq j \leq k\}$ such that

$$a(Z_{\alpha_j}) = e^{2\pi i \theta_j} Z_{\alpha_j}, \quad 1 \leq j \leq k.$$

Set $H \in \mathfrak{h}$ be the element such that

$$(H, \alpha_j) = -\theta_j, \quad 1 \leq j \leq k.$$

Then $\mathrm{Exp}\, H \in T$ and

$$\begin{cases} \mathrm{Exp}\, H \cdot a|C_0 = \mathrm{Id}_{C_0}, \\ \mathrm{Exp}\, H \cdot a(Z_{\alpha_j}) = Z_{\alpha_j}, \quad \mathrm{Exp}\, H \cdot a(Z_{-\alpha_j}) = Z_{-\alpha_j}. \end{cases}$$

Since $\{\mathfrak{h}, Z_{\alpha_j}, Z_{-\alpha_j}, 1 \leq j \leq k\}$ already generates $\mathfrak{G} \otimes \mathbb{C}$, $\mathrm{Exp}\, H \cdot a = \mathrm{Id}$, i.e. $a = \mathrm{Exp}(-H) \in T$. This proves the injectivity of ρ and hence the isomorphism

$$\mathrm{Aut}(\mathfrak{G})/\mathrm{Ad}(\mathfrak{G}) \cong \mathrm{Isom}(\pi). \qquad \square$$

2. Classification of Geometric Root Patterns

Theorem 1 effectively reduces the classification of simple compact Lie algebra to that of the homothety-types of their simple root systems.

Lemma 1. *A compact Lie algebra $\mathfrak{G}$ is simple if and only if its simple root system π spans $\mathfrak{h}$ and has no non-trivial splitting into mutually orthogonal subsets.*

Proof: It is easy to see that $\mathfrak{G}$ is semi-simple if and only if π spans $\mathfrak{h}$. If $\mathfrak{G}$ is semi-simple but non-simple, then

$$\mathfrak{G} = \mathfrak{G}_1 \oplus \cdots \oplus \mathfrak{G}_l, \quad \pi(\mathfrak{G}) = \pi(\mathfrak{G}_1) \oplus \cdots \oplus \pi(\mathfrak{G}_l),$$

where $\pi(\mathfrak{G}_j)$, $1 \leq j \leq l$ are mutually orthogonal. Conversely, suppose π splits into two mutually orthogonal non-trivial subsets, namely

$$\pi = \pi' \cup \pi'', \quad \pi' \perp \pi''.$$

Let $\alpha \in \pi'$, $\beta \in \pi''$ and $\mathfrak{h}_{\alpha\beta} = \langle \alpha, \beta \rangle^{\perp}$, $T_{\alpha\beta}$ to be the subtorus with $\mathfrak{h}_{\alpha\beta}$ as its Lie algebra. Let $G_{\alpha\beta}$ be the centralizer of $T_{\alpha\beta}$ and $\tilde{G}_{\alpha\beta} = G_{\alpha\beta}/T_{\alpha\beta}$. Then $\tilde{G}_{\alpha\beta}$ is a compact connected Lie group of rank 2 and $\Delta(\tilde{G}_{\alpha\beta}) = \{\pm\alpha, \pm\beta\}$. Therefore, $\tilde{G}_{\alpha\beta}$ is covered by $S^3 \times S^3$ and hence

$$[\mathbb{R}^2_{(\pm\alpha)}, \mathbb{R}^2_{(\pm\beta)}] = 0,$$

for $\alpha \in \pi'$ and $\beta \in \pi''$. Set $\mathfrak{G}'$ and $\mathfrak{G}''$ to be the subalgebras generated by

$$\{\mathbb{R}^2_{(\pm\alpha)}, \alpha \in \pi'\} \quad \text{and} \quad \{\mathbb{R}^2_{(\pm\beta)}, \beta \in \pi''\}$$

respectively. Then it is not difficult to see that $\mathfrak{G} = \mathfrak{G}' \oplus \mathfrak{G}''$ and hence non-simple. □

Schematically, it is convenient to record the angles between the simple roots $\alpha_j \in \pi$ by a diagram defined as follows.

(i) Each simple root is simply represented by a dot.

(ii) Two dots are joined by 0, 1, 2, or 3 lines according to the angle between the two corresponding roots, that is, $\pi/2$, $2\pi/3$, $3\pi/4$ or $5\pi/6$ (cf. Lemma 4.5).

Remarks (i) π is non-splitable if and only if its associated diagram is *connected*.

(ii) One may also consider the above diagram as the book-keeping device of the geometry of the Weyl chamber $\bar{C}_0$, namely, each dot denotes a wall and the number of lines joining two dots records the angle between the two walls, i.e., $\{0, 1, 2, 3\} \leftrightarrow \{\pi/2, \pi/3, \pi/4, \pi/6\}$.

The following is a simple classification result in the realm of elementary Euclidean geometry.

Theorem 3. *The following is a complete list of all geometrically feasible connected diagrams:*

(i) A_k: (*k dots*)

(ii) B_k or C_k: (*k dots, k ≥ 2*)

(iii) D_k: (*k dots, k ≥ 4*)

(iv) E_6:

E_7:

E_8:

(v) F_4:

(vi) G_2:

Remark   Two simple roots joined by a single bond are always of the same length. Two simple roots joined by a multiple bond are of different lengths, one usually adds a direction to the multiple bond to indicate that the latter one is shorter. In fact, only the second case will make an essential difference in this refined diagram, namely,

B_k:

C_k:

they are different for $k > 3$

Proof: The above theorem is a purely geometric fact in Euclidean space, namely, the possibilities of having k linearly independent unit vectors $\{e_j, 1 \leq j \leq k\}$ with specifically prescribed angles. We shall call a *geometrically realizable* diagram an *admissible* diagram. It is not difficult to see that such a set of unit vectors $\{e_j, 1 \leq j \leq k\}$ exists if and only if

$$\left| \sum_{j=1}^{k} x_j e_j \right|^2 = \sum_{i,j=1}^{k} x_i x_j (e_i, e_j) \geq 0,$$

and is equal to zero only when all $x_j = 0$. By taking special values of x_j, it is easy to obtain the following necessary conditions on the admissible diagrams.

(1) Subdiagrams of an admissible diagram are still admissible.

(2) An admissible diagram contains at most $(k-1)$ bonds.

Proof:

$$\left| \sum_{j=1}^{k} \mathbf{e}_j \right|^2 = k + 2 \sum_{i<j} (\mathbf{e}_i, \mathbf{e}_j) > 0$$

implies that the number of non-zero terms in $(\mathbf{e}_i, \mathbf{e}_j)$ is at most $(k-1)$.

(3) (1) and (2) imply that there is no cycles in the diagram.

(4) No more than three lines can be joined to a single dot.

Proof: Suppose $\mathbf{e}_1, \ldots, \mathbf{e}_l$ are joined to $\mathbf{e}_{l+1}$. Then by (3), $\mathbf{e}_1, \ldots, \mathbf{e}_l$ are orthonormal. Extend them to an orthonormal basis of $\langle \mathbf{e}_1, \ldots, \mathbf{e}_l, \mathbf{e}_{l+1} \rangle$ by adding $\tilde{\mathbf{e}}_{l+1}$. Then

$$\mathbf{e}_{l+1} = \sum_{i=1}^{l} (\mathbf{e}_i, \mathbf{e}_{l+1}) \cdot \mathbf{e}_i + (\tilde{\mathbf{e}}_{l+1}, \mathbf{e}_{l+1}) \tilde{\mathbf{e}}_{l+1}.$$

Hence

$$\sum (\mathbf{e}_i, \mathbf{e}_{l+1})^2 = 1 - (\tilde{\mathbf{e}}_{l+1}, \mathbf{e}_{l+1})^2 < 1,$$

namely, $\sum 4(\mathbf{e}_i, \mathbf{e}_{l+1})^2 < 4$. □

(5) The diagram obtained by contracting a subdiagram of the type ∘—∘⋯∘—∘ in an admissible diagram to a single dot is still admissible.

Proof: Suppose $\mathbf{e}_1, \ldots, \mathbf{e}_l$ are the unit vectors with a subdiagram of the above type. Then $\mathbf{e}_1 + \cdots + \mathbf{e}_l$ is again a unit vector and the diagram of

$$\{(\mathbf{e}_1 + \cdots + \mathbf{e}_l), \mathbf{e}_{l+1}, \ldots, \mathbf{e}_k\}$$

is exactly the contracted diagram. [It is easy to see the case $l = 2$.]

It follows easily from (1)–(5) that

(i) ∘━━∘ is the only admissible diagram with a triple bond.

(ii) An admissible diagram contains no subdiagram of the following type, namely,

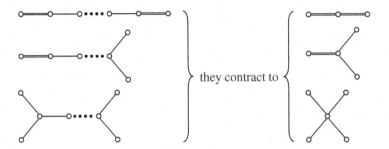

and hence contradicts (4).

Finally, let us determine which diagrams of the following type are admissible, namely,

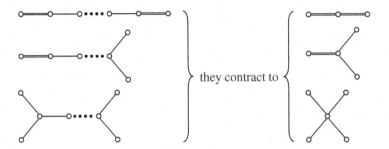

$$p \geq q \geq 1,$$

$$p \geq q \geq r \geq 2.$$

Set

$$\tilde{\mathbf{e}} = \sum_{i=1}^{p} i\mathbf{e}_i, \quad \tilde{\mathbf{f}} = \sum_{j=1}^{q} j\mathbf{f}_j,$$

$$\tilde{\mathbf{b}} = \sum_{i=1}^{p-1} i\mathbf{b}_i, \quad \tilde{\mathbf{c}} = \sum_{i=1}^{q-1} i\mathbf{c}_i, \quad \tilde{\mathbf{d}} = \sum_{i=1}^{r-1} i\mathbf{d}_i.$$

Then straightforward computations will show that

(i)
$$\begin{cases} (\tilde{\mathbf{e}}, \tilde{\mathbf{e}}) = \dfrac{p(p+1)}{2}, (\tilde{\mathbf{f}}, \tilde{\mathbf{f}}) = \dfrac{q(q+1)}{2}, \\[2mm] (\tilde{\mathbf{e}}, \tilde{\mathbf{f}})^2 = p^2 q^2 (\mathbf{e}_p, \mathbf{f}_q)^2 = \dfrac{p^2 q^2}{2}, \\[2mm] (\tilde{\mathbf{e}}, \tilde{\mathbf{f}})^2 < (\tilde{\mathbf{e}}, \tilde{\mathbf{e}}) \cdot (\tilde{\mathbf{f}}, \tilde{\mathbf{f}}) \Rightarrow (p-1)(q-1) < 2 \Rightarrow \begin{cases} (p, q) = (2, 2) \\ q = 1 \end{cases}, \end{cases}$$

(ii)
$$
\begin{cases}
(\tilde{\mathbf{b}}, \tilde{\mathbf{b}}) = \dfrac{1}{2}p(p-1), (\tilde{\mathbf{c}}, \tilde{\mathbf{c}}) = \dfrac{1}{2}q(q-1), (\tilde{\mathbf{d}}, \tilde{\mathbf{d}}) = \dfrac{1}{2}r(r-1), \\[2mm]
\left.\begin{array}{l}
(\tilde{\mathbf{b}}, \mathbf{a})^2/(\tilde{\mathbf{b}}, \tilde{\mathbf{b}}) = \dfrac{1}{2}\left(1 - \dfrac{1}{p}\right) \\[2mm]
(\tilde{\mathbf{c}}, \mathbf{a})^2/(\tilde{\mathbf{c}}, \tilde{\mathbf{c}}) = \dfrac{1}{2}\left(1 - \dfrac{1}{q}\right) \\[2mm]
(\tilde{\mathbf{d}}, \mathbf{a})^2/(\tilde{\mathbf{d}}, \tilde{\mathbf{d}}) = \dfrac{1}{2}\left(1 - \dfrac{1}{r}\right)
\end{array}\right\} \Rightarrow \dfrac{1}{p} + \dfrac{1}{q} + \dfrac{1}{r} > 1
\end{cases}
$$

$$
\Rightarrow \begin{cases}
q = r = 2, \quad p \text{ arbitrary} \\
r = 2, q = 3, 3 \le p \le 5.
\end{cases}
$$

Summarizing the above rather elementary detail discussions, one shows that the diagrams listed in Theorem 3 are, indeed, the only admissible connected diagrams. It is not difficult to construct an explicitly given set of unit vectors to demonstrate that all of them are geometrically realizable. $\qquad\square$

Remark   Of course, it is a problem of different order of magnitude to determine whether they can all be realized as the diagram for the simple root system of compact Lie algebras. However, the remarkable final results of the classification theory of compact Lie algebras is exactly that each one of the above diagram can be realized as the diagram of π for a *unique* simple, compact Lie algebra (up to isomorphism)!

Exercises   1. Construct explicit sets of unit vectors whose diagrams are exactly A_k, B_k, D_k, E_6, E_7, E_8, F_4, G_2, respectively.

2. Show that an isometry of two simple root systems uniquely extends to an isometry of the root system.

3. Classical Compact Lie Groups and Their Root Systems

(I) $U(n)$ and $SU(n)$

Let $U(n)$ be the group of $n \times n$ unitary matrices and $SU(n)$ be the subgroup of $n \times n$ unitary matrices with determinant 1. Let μ_n be the representation of $U(n)$ on $\mathbb{C}^n \simeq M_{n,1}(\mathbb{C})$ via matrix multiplication and $\mu'_n = \mu_n | \mathrm{SU}(n)$. Then

$$T^n = \{\mathrm{diag}(e^{2\pi i \theta_1}, \ldots, e^{2\pi i \theta_n}), 0 \le \theta_j \le 1\}$$

is a maximal torus of $U(n)$ and the subtorus $T^{n-1} \subset T^n$ defined by the condition $\theta'_1 + \theta'_2 + \cdots + \theta'_n = 0$ is a maximal torus of $\mathrm{SU}(n)$. Hence

$$\Omega(\mu_n) = \{\theta_j; 1 \le j \le n\},$$

$$\Omega(\mu'_n) = \{\theta'_j; 1 \le j \le n\}, \quad \sum \theta'_j = 0.$$

The Lie algebra of $U(n)$ consists of all skew hermitian matrices and hence its complexification of $\mathfrak{gl}(n, \mathbb{C}) = M_{n,n}(\mathbb{C})$. Therefore,

$$\mathrm{Ad}_{U(n)} \otimes \mathbb{C} = \mu_n \otimes_\mathbb{C} \mu_n^*,$$

$$\mathrm{Ad}_{\mathrm{SU}(n)} \otimes \mathbb{C} = \mu'_n \otimes_\mathbb{C} \mu_n'^* - 1,$$

and hence

$$\Delta(U(n)) = \{(\theta_j - \theta_k), 1 \le j \ne k \le n\},$$

$$\Delta(\mathrm{SU}(n)) = \{(\theta'_j - \theta'_k), 1 \le j \ne k \le n\}, \sum \theta'_j = 0.$$

It is quite natural to choose the ordering such that

$$\theta'_1 > \theta'_2 > \cdots > \theta'_{n-1} \quad \text{(for the case } \mathrm{SU}(n)\text{)}.$$

Then

$$\Delta^+(\mathrm{SU}(n)) = \{(\theta'_j - \theta'_k), 1 \le j < k \le n\},$$

and

$$\pi = \{\theta'_j - \theta'_{j+1}, 1 \le j \le n-1\}.$$

The Weyl group acts on

$$\mathfrak{h} = \left\{ (\theta'_1, \theta'_2, \ldots, \theta'_n), \sum \theta'_j = 0 \right\}$$

as the permutations of the n coordinates. It is convenient to regard $\theta'_j = \theta_j - \frac{1}{n} \sum \theta_i$ where $\{\theta_i; 1 \le i \le n\}$ is an orthonormal basis. Therefore, one has the following diagram.

$$\theta'_1 - \theta'_2 \quad \theta'_2 - \theta'_3 \quad \theta'_j - \theta'_{j+1} \quad \theta'_{n-1} - \theta'_n \qquad (A_{n-1})$$

(II) SO(n)

Let $O(n)$ be the group of $n \times n$ orthogonal matrices. It consists of two connected components with determinants of ± 1 respectively; SO(n) is the subgroup of $n \times n$ orthogonal matrices of determinant 1. Let ρ_n be the representation of SO(n) on $\mathbb{R}^n = M_{n,1}(\mathbb{R})$ via matrix multiplication. Then

$$\mathrm{Ad}_{\mathrm{SO}(n)} = \Lambda^2 \rho_n,$$

namely, the conjugation of anti-symmetric matrices by orthogonal matrices. It is convenient to choose the maximal torus of SO(n) as follows.

$$n = 2k \colon T^k = \left\{ \begin{pmatrix} \boxed{\begin{matrix} \cos 2\pi\theta_1 & -\sin 2\pi\theta_1 \\ \sin 2\pi\theta_1 & \cos 2\pi\theta_1 \end{matrix}} & & \\ & \ddots & \\ & & \boxed{\begin{matrix} \cos 2\pi\theta_k & -\sin 2\pi\theta_k \\ \sin 2\pi\theta_k & \cos 2\pi\theta_k \end{matrix}} \end{pmatrix} \right\},$$

$n = 2k+1$:

$$T^k = \left\{ \begin{pmatrix} \boxed{\begin{matrix} \cos 2\pi\theta_1 & -\sin 2\pi\theta_1 \\ \sin 2\pi\theta_1 & \cos 2\pi\theta_1 \end{matrix}} & & \\ & \ddots & \\ & & \boxed{\begin{matrix} \cos 2\pi\theta_k & -\sin 2\pi\theta_k \\ \sin 2\pi\theta_k & \cos 2\pi\theta_k \end{matrix}} \\ & & & 1 \end{pmatrix} \right\}.$$

Then

$$\Omega(\rho_{2k}) = \{\pm\theta_j, 1 \le j \le k\},$$
$$\Omega(\rho_{2k+1}) = \{\pm\theta_j, 1 \le j \le k; 0\}.$$

Therefore

$$\Delta(\mathrm{SO}(2k)) = \{\pm\theta_i \pm \theta_j, 1 \le i \le j \le k\},$$
$$\Delta(\mathrm{SO}(2k+1)) = \{\pm\theta_i \pm \theta_j, 1 \le i < j \le k; \pm\theta_i, 1 \le i \le k\},$$

and hence the Weyl group action on $\mathfrak{h} = \{(\theta_1, \dots, \theta_k)\}$ is as follows.

$W(\mathrm{SO}(2k))$: permutations with even number of sign-changings.

$W(\mathrm{SO}(2k+1))$: permutations with arbitrary sign-changings.

It is convenient to fix the ordering such that

$$\theta_1 > \theta_2 > \cdots > \theta_k$$

and regard $\{\theta_i; 1 \leq i \leq k\}$ as an orthonormal basis.

Therefore,

$$\Delta^+(SO(2k)) = \{\theta_i \pm \theta_j, 1 \leq i < j \leq k\},$$

$$\Delta^+(SO(2k+1)) = \{\theta_i \pm \theta_j, 1 \leq i < j \leq k; \theta_i, 1 \leq i \leq k\},$$

$$\pi(SO(2k)) = \{\theta_j - \theta_{j+1}, 1 \leq j \leq k - 1; \theta_{k-1} + \theta_k\},$$

$$\pi(SO(2k+1)) = \{\theta_j - \theta_{j+1}, 1 \leq j \leq k - 1; \theta_k\}.$$

Hence one has the following diagrams.

$SO(2k)$, $k \geq 4$:

$SO(2k+1)$, $k \geq 2$:

Remark   The diagram of $SO(6)$ is the same as $SU(4)$. In fact, this implies that the above two groups are locally isomorphic. Actually, $\Lambda^2 \mu_4 : SU(4) \rightarrow SO(6) \subset SU(6)$.

(III) Sp(n) (The symplectic group of rank n)

Let $\mathbf{H}$ be the skew field of quaternions and

$$\mathbf{H}^n = \{(q_1, q_2, \ldots, q_n) : q_j \in \mathbb{H}\}$$

be the right free $\mathbf{H}$-module of rank n. We shall equip it with the following hermitian product:

$$\langle (q_1, q_2, \ldots, q_n), (q_1', q_2', \ldots, q_n') \rangle := \sum_{j=1}^{n} \bar{q}_j q_j'.$$

Then, the group of all isometries of $\mathbf{H}^n$ is called the symplectic group of rank n and shall be denoted by $Sp(n)$.

Examples 1. Sp(1) is exactly the multiplicative group of unit quaternions, acting on $\mathbf{H}^1$ via left multiplications, namely,

$$\mathrm{Sp}(1) = S^3 = \{q \in \mathbf{H}; q\bar{q} = 1\},$$

and $S^3 \times \mathbf{H} \to \mathbf{H}$ is given by $(q, q_1) = q \cdot q_1$.

2. Let $\mathbf{e}_1 = (1, 0, \ldots, 0), \ldots, \mathbf{e}_j = (0, \ldots, 1, \ldots, 0), \ldots, \mathbf{e}_n = (0, \ldots, 1)$ and g be an arbitrary element of $\mathrm{Sp}(n)$. Then $\{\mathbf{b}_j = g(\mathbf{e}_j), 1 \leq j \leq n\}$ is clearly an orthonormal basis of $\mathbf{H}^n$, namely

$$\langle \mathbf{b}_i, \mathbf{b}_j \rangle = \langle g\mathbf{e}_i, g\mathbf{e}_j \rangle = \langle \mathbf{e}_i, \mathbf{e}_j \rangle = \delta_{ij}.$$

Conversely, let $\{\mathbf{b}_j, 1 \leq j \leq n\}$ be an arbitrary orthonormal basis in $\mathbf{H}^n$. Then there exists a unique element $g \in \mathrm{Sp}(n)$ with $g(\mathbf{e}_j) = \mathbf{b}_j$, $1 \leq j \leq n$. Furthermore, it follows from the usual Gram–Schmidt orthogonalization that any unit vector $\mathbf{b}_1 \in \mathbf{H}^n$ can be extended to an orthonormal basis $\{\mathbf{b}_1, \mathbf{b}_2, \ldots, \mathbf{b}_n\}$ in $\mathbf{H}^n$. Therefore, $\mathrm{Sp}(n)$ acts transitively on the unit sphere, namely

$$S^{4n-1} = \left\{ \mathbf{u} = (q_1, \ldots, q_n); |\mathbf{u}|^2 = \sum_{j=1}^{n} \bar{q}_i q_j = 1 \right\}.$$

3. Let G_{e_n} be the subgroup of $\mathrm{Sp}(n)$ which fixes $\mathbf{e}_n$. Then it is clear that $G_{e_n} \simeq \mathrm{Sp}(n-1)$. Therefore

$$S^{4n-1} = G(e_n) \cong \mathrm{Sp}(n)/\mathrm{Sp}(n-1),$$

and hence

$$\dim\mathrm{Sp}(n) = \dim\mathrm{Sp}(n-1) + (4n-1)$$

$$= \dim\mathrm{Sp}(n-2) + (4n-5) + (4n-1)$$

$$= \sum_{j=1}^{n}(4j-1) = \frac{1}{2}n(4n+2) = 2n^2 + n.$$

4. One may also consider $\mathbf{H}^n$ as a right $\mathbb{C}$-module of rank $2n$, namely, identifying $(q_1, q_2, \ldots, q_n)$ with

$$(u_1, u_2, \ldots, u_n; v_1, v_2, \ldots, v_n), \quad q_l = u_l + jv_l, \quad 1 \leq l \leq n.$$

Then $\mathrm{Sp}(n)$ is a subgroup of $U(2n)$ leaving a non-degenerate skew symmetric form invariant. We shall denote the above representation of $\mathrm{Sp}(n)$ on $\mathbb{C}^{2n}$ by ν_n (cf. Ch I. Chavalley's Lie group Theory).

Lemma 2. $\nu_n = \nu_n^*$ *and* $\mathrm{Ad}_{\mathrm{Sp}(n)} \otimes \mathbb{C} = S^2 \nu_n$.

Proof: Since $\mathrm{Sp}(n)$ acts transitively over the unit sphere of $\mathbb{C}^{2n}$, ν_n is clearly an irreducible representation.

$$\nu_n \otimes \nu_n = \Lambda^2 \nu_n \oplus S^2 \nu_n,$$

and $\Lambda^2 \nu_n$ contains a trivial copy because $\mathrm{Sp}(n)$ keeps a skew symmetric form invariant. Hence, it follows from the Schur lemma that $\nu_n = \nu_n^*$.

$\mathrm{Sp}(n)$ is a subgroup of $U(2n)$ and $\nu_n = \mu_{2n}|\mathrm{Sp}(n)$. Hence $\mathrm{Ad}_{\mathrm{Sp}(n)} \otimes \mathbb{C}$ is a component of $(\mathrm{Ad}_{U(2n)}|\mathrm{Sp}(n)) \otimes \mathbb{C}$, and

$$(\mathrm{Ad}_{U(2n)} \otimes \mathbb{C})|\mathrm{Sp}(n) = \nu_n \otimes \nu_n^* = \nu_n \otimes \nu_n$$

$$= \Lambda^2 \nu_n \oplus S^2 \nu_n,$$

where $\dim S^2 \nu_n = 2n^2 + n$, $\dim \Lambda^2 \nu_n = 2n^2 - n$. Therefore, the *irreducibility* of $\mathrm{Ad}_{\mathrm{Sp}(n)} \otimes \mathbb{C}$ will certainly imply that $\mathrm{Ad}_{\mathrm{Sp}(n)} \otimes \mathbb{C} = S^2(\nu_n)$. We shall prove the irreducibility of $\mathrm{Ad}_{\mathrm{Sp}(n)} \otimes \mathbb{C}$ by induction on n as follows. The case $n = 1$ is simple and well-known. Let us begin with the case $n = 2$. Recall that $\mathrm{Sp}(2)$ is a subgroup of $SU(4)$ and

  (i) $\Lambda^2 \mu_4 : SU(4) \to SO(6) \subset U(6)$,
 (ii) $\Lambda^2 \mu_4|\mathrm{Sp}(2) = \Lambda^2 \nu_2$ contains a trivial copy,
(iii) $\dim \mathrm{Sp}(2) = 10 = \dim SO(5)$.

Therefore, $(\Lambda^2 \nu_2 - 1) : \mathrm{Sp}(2) \to SO(5)$ is a covering homomorphism and hence $\mathrm{Ad}_{\mathrm{Sp}(2)} \otimes \mathbb{C}$ is irreducible because $\mathrm{Ad}_{SO(5)} \otimes \mathbb{C}$ is already known to be irreducible.

The general case $n \geq 3$: Let $\mathrm{Sp}(n-1)^{(j)}$ be the subgroup of $\mathrm{Sp}(n)$ which fixes $\mathbf{e}_j = (0, \ldots, 1, \ldots, 0)$ and $\mathfrak{G}^{(j)}$ be the Lie subalgebra of $\mathrm{Sp}(n-1)^{(j)}$. By the induction assumption that $\mathrm{Ad}_{\mathrm{Sp}(n-1)} \otimes \mathbb{C}$ is irreducible, each $\mathfrak{G}^{(j)} \otimes \mathbb{C}$ must be contained in an irreducible subspace of $\mathfrak{G} \otimes \mathbb{C}$ say V_j, $1 \leq j \leq n$. Therefore

$$V_j \cap V_l \supset (\mathfrak{G}^{(j)} \cap \mathfrak{G}^{(l)}) \otimes \mathbb{C} \neq \{0\},$$

for all $1 \leq j, l \leq n$ and hence $V_j = V_l$ for all $1 \leq j, l \leq n$, namely, $V_j = \mathfrak{G} \otimes \mathbb{C}$ and hence $\mathrm{Ad}_{\mathrm{Sp}(n)} \otimes \mathbb{C}$ is irreducible. $\qquad\square$

Let $T^n = U(1)^n \subset (\mathrm{Sp}(1))^n \subset \mathrm{Sp}(n)$. Then

$$\Omega(\nu_n | T^n) = \{\pm\theta_i, 1 \le i \le n\},$$

$$\Omega(S^2 \nu_n | T^n) = \{\pm\theta_i \pm \theta_j, 1 \le i < j \le n; \pm 2\theta_i, 1 \le i \le n \quad \text{and } 0$$
$$\text{with multi. } n\}.$$

This shows that T^n is in fact a maximal torus of $\mathrm{Sp}(n)$ and

$$\Delta(\mathrm{Sp}(n)) = \{\pm\theta_i \pm \theta_j, 1 \le i < j \le n; \pm 2\theta_i, 1 \le i \le n\}.$$

Again, it is convenient to fix the ordering such that

$$\theta_1 > \theta_2 > \cdots > \theta_n$$

and regard $\{\theta_i, 1 \le i \le n\}$ as an orthonormal basis of $\mathfrak{h}$. [The Weyl group is, in fact, isomorphic to that of $\mathrm{SO}(2n+1)$ as a transformation group.] Therefore

$$\Delta^+(\mathrm{Sp}(n)) = \{\theta_i \pm \theta_i, 1 \le i < j \le n; 2\theta_i, 1 \le i \le n\},$$

$$\pi = \{\theta_i - \theta_{i+1}, 1 \le i \le n-1; 2\theta_n\},$$

and one has the following diagram of C_n-type.

$$\theta_1 - \theta_2 \quad \theta_2 - \theta_3 \quad \theta_{n-i} - \theta_n \quad 2\theta_n \qquad (C_n)$$

Lecture 7

Basic Structural Theory of Lie Algebras

In this lecture, we shall study the basic structural theory of Lie algebras in general, namely, beyond the rather special case of *compact Lie algebras*. For example, the set of linear transformations of a vector space V over $\Bbbk$ constitutes a Lie algebra with $[A, B] = AB - BA$, denoted by $\mathfrak{gl}(V)$ or $\mathfrak{gl}(n, \Bbbk)$, while a sub-Lie algebra $\mathfrak{g} \subset \mathfrak{gl}(V)$ will be regarded as a linear representation of $\mathfrak{g}$ on V. In the special, but of basic importance, case of $\Bbbk = \mathbb{C}$, one has the well-known theorem of Jordan canonical form and Jordan decomposition for a *single* linear transformation A (cf. Section 1.1). It is natural to seek ways of generalizing such a wonderful result beyond a single linear transformation, thus asking what would be a proper setting for such a generalization? As one shall find out in the discussion of this lecture, representations of Lie algebras will provide an advantageous setting for such a task and it naturally leads to the basic structure theory of Lie algebras.

1. Jordan Decomposition: A Review and an Overview

1.1. *Jordan decomposition*

Let V be a complex vector space and $A \in \mathfrak{gl}(V)$ a given linear transformation on V. One has the following remarkable theorem, namely,

Theorem 1 (Jordan). *There exists a unique decomposition of A into the sum of* semi-simple (*resp.* nilpotent) *linear transformations satisfying the following properties, namely,*

$$A = s(A) + n(A) \quad and \quad [s(A) + n(A)] = 0$$

where $n(A)$ is nilpotent and $s(A)$ is semi-simple (i.e. its minimal polynomial has no multiple roots), while V has the following direct sum decomposition of A-invariant subspaces, namely,

$$V = \bigoplus_{i=1}^{l} V_i, \quad V_i = \ker(A - \lambda_i I)^{\alpha_i}.$$

Proof: (sketchy) First of all, the following seemingly trivial fact plays a crucial role, namely,

$$[A, B] = 0 \Longrightarrow \ker B \text{ is } A\text{-invariant}$$

$$(\text{i.e. } BX = 0 \Longrightarrow BAX = ABX = 0).$$

On the other hand, to any polynomial $f(x) \in \mathbb{C}[x]$, $f(A)$ is obviously commuting with A, thus bringing the well-known theorems on complex coefficient polynomials as a powerful tool in analyzing the decomposition of V into a direct sum of A-invariant subspaces. Let

$$m_A(x) = \prod_{i=1}^{l}(x - \lambda_i)^{\alpha_i}$$

be the minimal polynomial of A, where $\{\lambda_i\}$ are its distinct roots (i.e. eigenvalues). Then

$$V = \bigoplus_{i=1}^{l} V_i, \quad V_i = \ker(A - \lambda_i I)^{\alpha_i}.$$

Set

$$f_i(x) = \frac{m_A(x)}{(x - \lambda_i)^{\alpha_i}} = \prod_{j \neq i}(x - \lambda_j)^{\alpha_j}.$$

For $\lambda_i \neq 0, x f_i(x)$ and $(x - \lambda_i)^{\alpha_i}$ are relatively prime. Therefore, there exists $g_i(x), h_i(x)$ such that

$$g_i(x) x f_i(x) + h_i(x)(x - \lambda_i)^{\alpha_i} = \lambda_i.$$

Set

$$S_A(x) = \sum_{i=1}^{l} g_i(x) x f_i(x), N_A(x) = x - S_A(x).$$

Then, it is quite straightforward to verify the following:

(i) $S_A(0) = N_A(0) = 0$,

(ii) $A = S_A(A) + N_A(A)$, namely, $S_A(x) + N_A(x) \equiv x \pmod{m_A(x)}$,

(iii) $N_A(A)$ is nilpotent and $S_A(A)$ is semi-simple (i.e. its minimal polynomial has no multiple root),

(iv) suppose $A = S + N, [S, N] = 0$, and S is semi-simple and N is nilpotent. Then $S = S_A(A)$ and $N = N_A(A)$.   □

Remark A simple modification of the proof will also show the unique existence of another semi-simple element, say denoted by $\bar{s}(A)$, such that

$$\bar{s}(A)|_{V_i} = \bar{\lambda}_i I, \quad 1 \leq i \leq l.$$

Namely, let $\hat{g}_i(x), \hat{h}_i(x)$ be a pair of polynomials such that

$$\hat{g}_i(x) x f_i(x) + \hat{h}_i(x)(x - \lambda_i)^{\alpha_i} = \bar{\lambda}_i.$$

Then one has $\bar{s}(A) = \bar{S}_A(x)$, where

$$\bar{S}_A(x) = \sum_{i=1}^{l} \hat{g}_i(x) x f_i(x).$$

Thus, one has the following useful lemma, namely,

Lemma 1. $A \in \mathfrak{gl}(V)$ *is nilpotent if and only if* $\mathrm{Tr}(A\bar{s}(A)) = 0$.

Proof: $\mathrm{Tr}(A\bar{s}(A)) = \sum_{i=1}^{l} \dim V_i |\lambda_i|^2 = 0$ holds if and only if $l = 1$ and $\lambda_1 = 0$, i.e., A is nilpotent.   □

Remarks (i) The Jordan canonical form is a refinement of the above theorem, which amounts to prove a simple additional lemma on the canonical form of nilpotent linear transformations which, in fact, holds for arbitrary coefficient field $\Bbbk$.

(ii) Theorem 1 itself can also be generalized to the case of an arbitrary field $\Bbbk$, and essentially by the same method of proof, namely,

$$V = \bigoplus_{i=1}^{l} V_i, \quad V_i = \ker(p_i(A)^{\alpha_i})$$

where

$$m_A(x) = \prod_{i=1}^{l} p_i(x)^{\alpha_i}$$

is the primary factorization of $m_A(x)$.

(iii) For a given collection of mutually commutative linear transformations, say $\{A_i\}$, it is quite straightforward to show the same kind of decomposition of V into *common* invariant subspace. Here, the commutativity is crucial and the seemingly trivial fact that we start with in the proof of Theorem 1 again plays the major rule.

1.2. *A review and an overview*

An analytical review of the result of Section 1.1 from the geometric point of view of linear transformations indicates that the central topic of the kind of generalization that we are seeking for *non-commutative collections* of linear transformations should be the *structure of common invariant subspaces*, while the special case of commutative collections makes such an analysis much easier (cf. Section 1.1). Let $\{A_j\}$ be such a collection and $\mathfrak{g}$ be the sub-Lie algebra generated by $\{A_j\}$. Then it is obvious that a $\{A_j\}$-invariant subspace is also a $\mathfrak{g}$-invariant subspace, while the operation $[.,.]$ of $\mathfrak{g}$ is exactly a systematic total accounting of the non-commutativity of $\{A_j\}$. Therefore, the representation theory of Lie algebras should be a proper setting for studying the structure of common invariant subspaces, because the Lie algebra structure *highlights the non-commutativity* and also provides various kinds of measurements of the degree of non-commutativity, such as nilpotency, solvability and semi-simplicity.

First, we shall give a brief overview of basic structural theory of Lie algebras.

1.2.1. *Basic definitions*

(i) A subspace $\mathfrak{a} \subseteq \mathfrak{g}$ is called an *ideal* of $\mathfrak{g}$ if $[\mathfrak{g}, \mathfrak{a}] \subseteq \mathfrak{a}$.

(ii) If $\mathfrak{a}$ is an ideal, then $[\mathfrak{g}, \mathfrak{a}]$ is also an ideal. One has

$$[\mathfrak{g}, [\mathfrak{g}, \mathfrak{a}]] \subset [[\mathfrak{g}, \mathfrak{g}], \mathfrak{a}]] + [\mathfrak{g}, [\mathfrak{g}, \mathfrak{a}]] \subset [\mathfrak{g}, \mathfrak{a}] + [\mathfrak{g}, \mathfrak{a}] = [\mathfrak{g}, \mathfrak{a}].$$

(iii) Set

$$D\mathfrak{g} = [\mathfrak{g}, \mathfrak{g}], \quad D^{n+1}\mathfrak{g} = [D^n\mathfrak{g}, D^n\mathfrak{g}]$$
$$C^1\mathfrak{g} = \mathfrak{g}, \quad C^{n+1}\mathfrak{g} = [\mathfrak{g}, C^n\mathfrak{g}].$$

Definition A Lie algebra $\mathfrak{g}$ is called a *nilpotent* (resp. *solvable*) Lie algebra, if there exists an n such that $C^n\mathfrak{g} = \{0\}$ (resp. $D^n\mathfrak{g} = \{0\}$).

(iv) Suppose $\mathfrak{a}$ and $\mathfrak{b}$ are both solvable ideals of $\mathfrak{g}$. Then it is easy to verify that $\mathfrak{a} + \mathfrak{b}$ is also a solvable ideal of $\mathfrak{g}$. Therefore, there is a *maximal solvable ideal* of $\mathfrak{g}$ containing all the others. It is defined to be the *radical* of $\mathfrak{g}$.

Definition A Lie algebra $\mathfrak{g}$ is defined to be *semi-simple* if its radical is trivial, i.e. $\{0\}$.

In general, the *quotient* Lie algebra of $\mathfrak{g}$ by its radical, say denoted by $\mathfrak{g}/\mathfrak{r}$, is semi-simple. Therefore, a Lie algebra, in general, is an extension of a semi-simple one by a solvable one, namely, one has the following exact sequence

$$0 \to \mathfrak{r} \to \mathfrak{g} \to \mathfrak{g}/\mathfrak{r} \to 0.$$

1.2.2. *Examples of solvable (resp. nilpotent) Lie algebras*

Example 1. Let V be a vector space of dimension n, and

$$\{0\} \subset V_1 \subset V_2 \subset \cdots V_i \subset \cdots \subset V_n = V, \quad \dim V_i = i,$$

be a sequence of subspaces. Set $\mathfrak{g}$ to be a sub-Lie algebra of $\mathfrak{gl}(V)$ which keeps each of V_i invariant. Then $\mathfrak{g}$ is solvable.

Proof: Let $\{\mathbf{b}_j; 1 \le j \le n\}$ be a basis for V such that $\{\mathbf{b}_j; 1 \le j \le i\}$ is a basis for V_i. Then the matrices of $X \in \mathfrak{g}$ are upper triangular, namely,

$$X\mathbf{b}_i = \lambda_i(X)\mathbf{b}_i \pmod{V_{i-1}}, \quad 1 \le i \le n.$$

Therefore

$$[X, Y]\mathbf{b}_i \equiv 0 \pmod{V_{i-1}}, \quad 1 \le i \le n$$

and furthermore, it is easy to prove inductively that, for $\omega \in D^m \mathfrak{g}$, one has

$$\omega V_i \subseteq V_{i-m}, \quad 1 \leq i \leq n.$$

Hence, $D^n \mathfrak{g} = \{0\}$, i.e. $\mathfrak{g}$ is solvable. □

Example 2. Suppose that $\mathfrak{g}_0 \subset \mathfrak{gl}(V)$ consists only of elements with

$$X \mathbf{b}_i \equiv \lambda(X) \mathbf{b}_i \pmod{V_{i-1}}, \quad 1 \leq i \leq n,$$

namely, all the diagonal elements of its upper triangular matrices are equal to each other. Then $\mathfrak{g}_0$ is nilpotent.

Proof: The same kind of computation will provide an inductive verification that

$$\omega \in C^m \mathfrak{g} \Longrightarrow \omega V_i \subseteq V_{i-m+1}, \quad 1 \leq i \leq n.$$ □

Example 3. Suppose $\mathfrak{g} \subset \mathfrak{gl}(V)$ has a $\mathfrak{g}$-invariant direct sum decomposition, say

$$V = U_1 \oplus U_2 \oplus \cdots \oplus U_l$$

such that $\mathfrak{g}|_{U_j}, 1 \leq j \leq l$, are all of the kind of Example 2. Then $\mathfrak{g}$ is, of course, also nilpotent.

1.2.3. *Some remarkable generalizations of Theorem 1 in the setting of Lie algebras and their linear representations*

First of all, in the special but of basic importance case of $\Bbbk = \mathbb{C}$, one has the following remarkable results generalizing the Jordan theorem in the setting of linear Lie algebras and their structures of invariant subspaces, namely, the following two theorems.

Theorem 2 (Lie). *A solvable sub-Lie algebra of $\mathfrak{gl}(n, \mathbb{C})$ is necessarily of the kind of Example 1, or equivalently, $\mathfrak{g} \subset \mathfrak{gl}(n, \mathbb{C})$ is solvable if and only if $\mathfrak{g}$ is conjugate to a sub-Lie algebra of that of upper triangular matrices.*

Theorem 3. *Let $\mathfrak{g} \subset \mathfrak{gl}(V)$ be a nilpotent sub-Lie algebra. Then $\mathfrak{g}$ is necessarily an example of Example 3, namely, there exists a decomposition* *of V into $\mathfrak{g}$-invariant subspaces*

$$V = \bigoplus_{i=1}^{l} U_i,$$

with $\mathfrak{g}|_{U_i}$, $1 \leq i \leq l$, which are examples of the type of Example 2.

Moreover, Theorem 3 can be further generalized to the general case of *arbitrary* coefficient field $\Bbbk$ (cf. Theorem $\tilde{3}$), thus achieving a direct generalization of Theorem 1 from a single linear transformation to that of nilpotent linear Lie algebras over an arbitrary coefficient field $\Bbbk$.

As one shall see in the discussions of the following sections of this lecture, the study of invariant subspaces of linear representations of Lie algebras also provides a natural way of developing the basic structure theory of Lie algebras.

2. Nilpotent Lie Algebras and Solvable Lie Algebras

2.1. *Nilpotent Lie algebras*

2.1.1. *Theorem of Engel*

Let $\mathfrak{g} \subset \mathfrak{gl}(V)$, $\Bbbk$ arbitrary field. If all elements of $\mathfrak{g}$ are nilpotent, then there exists a common eigenvector $\boldsymbol{b}$ with zero eigenvalue, namely,

$$X\boldsymbol{b} = 0, \quad \forall X \in \mathfrak{g}.$$

Proof: We shall prove this by induction on dim $\mathfrak{g}$, while the beginning case of dim $\mathfrak{g} = 1$ is trivial and the induction assumption is that the above theorem holds for dim $\mathfrak{g} = m - 1$ in *arbitrary* $\mathfrak{gl}(V)$.

(i) Suppose dim $\mathfrak{g} = m$ and $\mathfrak{h}$ is a proper sub-Lie algebra of $\mathfrak{g}$. Set $W = \mathfrak{g}/\mathfrak{h}$ to be the *quotient space*. Let X to be a given element in $\mathfrak{h}$ and X^* be the induced linear transformation of $\mathrm{ad}_X \mathfrak{g}$. (Note that $\mathfrak{h}$ is, of course, invariant under ad_X.) On the other hand, one has

$$L_X \text{ (resp. } R_X, \mathrm{ad}_X \mathfrak{gl}(V)) \colon \mathfrak{gl}(V) \to \mathfrak{gl}(V)$$

$$L_X Y = XY, \; R_X Y = YX, \mathrm{ad}_X \mathfrak{gl}(V) = L_X - R_X$$

are commuting linear transformations on $\mathfrak{gl}(V)$ and L_X (resp. R_X) are by assumption nilpotent. Hence,

$$\mathrm{ad}_X \mathfrak{g} = \mathrm{ad}_X \mathfrak{gl}(V)|_{\mathfrak{g}}$$

and $X^* \in \mathfrak{gl}(W)$ are also nilpotent for $X \in \mathfrak{h}$. Thus

$$\mathfrak{h}^* = \{X^*; X \in \mathfrak{h}\}$$

is a sub-Lie algebra in $\mathfrak{gl}(W)$ with dim $\mathfrak{h}^* < m$. Hence, by induction assumption,

$$W_0 = \{w \in W; X^* w = 0, \forall X^* \in \mathfrak{h}^*\} \neq \{0\},$$

thus having

$$\mathfrak{g} \supseteq \mathfrak{g}_1 = \pi^{-1}(W_0) \supsetneq \mathfrak{h} = \pi^{-1}(0)$$

where $\mathfrak{g}_1$ is a sub-Lie algebra of $\mathfrak{g}$ and $\mathfrak{h}$ is an ideal of $\mathfrak{g}_1$.

(ii) Now, let $\mathfrak{g}'$ be a maximal proper sub-Lie algebra of $\mathfrak{g}$. By (i) $\mathfrak{g}'$ is an ideal of $\mathfrak{g}$ and moreover, $\mathfrak{g}/\mathfrak{g}'$ must be one-dimensional, lest $\mathfrak{g}'$ will not be maximal. Therefore, again by induction assumption,

$$V_0 = \{v \in V; Xv = 0, \ \forall X \in \mathfrak{g}'\} \neq \{0\}.$$

Note that

$$\mathfrak{g} = \mathfrak{g}' + \langle Y \rangle \neq \mathfrak{g}'$$

and V_0 is Y-invariant because

$$X(Yv) = Y(Xv) + [X, Y]v = 0, \ X \in \mathfrak{g}', \ v \in V_0.$$

By the assumption of the theorem, Y is nilpotent and hence $Y|_{V_0}$ is also nilpotent, thus having an eigenvector b in V_0 with zero eigenvalue, namely, a common eigenvector of $\mathfrak{g}$ with zero eigenvalue. $\qquad \square$

Corollary 1. *Such a $\mathfrak{g}$ is nilpotent and of the type of Example 2.*

Proof: A straightforward proof by induction on dim V. $\qquad \square$

Corollary 2. *A Lie algebra $\mathfrak{g}$ is nilpotent if and only if the operators ad_X are nilpotent for all $X \in \mathfrak{g}$.*

2.1.2. *Generalization of Jordan decomposition for nilpotent Lie algebras, primary decomposition*

For a single given $A \in \mathfrak{gl}(V)$, $\Bbbk$ arbitrary, let

$$m_A(X) = \prod_{i=1}^{l} p_i(X)^{\alpha_i}$$

where $p_i(X)$ are irreducible in $\Bbbk[X]$ and relatively prime to each other. Then V has a decomposition into A-invariant subspaces, namely,

$$V = \bigoplus_{i=1}^{l} V_i, \ \ V_i = \ker(p_i(X)^{\alpha_i}),$$

often referred to as the primary decomposition of A, which is a direct generalization of Theorem 1 in the special case of $\Bbbk = \mathbb{C}$.

In the setting of Lie algebra representations, the following theorem is a far-reaching direct generalization of primary decomposition for a single $A \in \mathfrak{gl}(V)$ to that of a nilpotent sub-Lie algebra $\mathfrak{g} \subset \mathfrak{gl}(V)$.

Theorem $\tilde{3}$. *Let $\mathfrak{g} \subset \mathfrak{gl}(V)$ be a nilpotent sub-Lie algebra. Then $\mathfrak{g}$ is necessarily an example of Example 3. Then there exists a decomposition of V into $\mathfrak{g}$-invariant subspaces*

$$V = \bigoplus_{i=1}^{l} V_i,$$

such that the minimal polynomial of $A_i = A|_{V_i}$ are primary for all $A \in \mathfrak{g}$ (i.e. having only one primary factor).

Proof: The important step of the proof is the following:

Let $A, B \in \mathfrak{g}$ and

$$m_A(x) = \prod_{i=1}^{l_A} p_i(x)^{\alpha_i}, \ \ \ V = \bigoplus_{i=1}^{l_A} V_i$$

be the primary decomposition of A. What we need to prove is that there exists an N such that

$$p_i(A)^N (B\boldsymbol{v}) = 0, \ \ \ \forall \boldsymbol{v} \in V_i.$$

We shall use the following identity of associative algebra: Let $\mathfrak{a}$ be an associative algebra and $a \in \mathfrak{a}$. Set

$$\mathrm{ad}_a : \mathfrak{a} \to \mathfrak{a}, \ \ \ \mathrm{ad}_a(b) = ab - ba$$

and hence
$$\mathrm{ad}_a(bc) = (ab - ba)c + b(ac - ca) = \mathrm{ad}_a(b)c + b\mathrm{ad}_a(c).$$

Therefore, if we use the shorthand notation denoting $\mathrm{ad}_a(*)$ just by $(*)'$ (i.e. view ad_a as a *derivation* operator on $\mathfrak{a}$), then one has the following identity, namely

$$a^k b = ba^k + \binom{k}{1} b'a^{k-1} + \binom{k}{2} b''a^{k-2} + \cdots + b^{(k)}.$$

Therefore, straightforward linear combination of the above identities for $k = 1, 2, \ldots$ will give the following identity for polynomials of a, namely,

$$f(a)b = bf(a) + b'f'(a) + b''\frac{1}{2!}f''(a) + \cdots + b^{(k)}\frac{1}{k!}f^{(k)}(a).$$

Now, since $\mathfrak{g}$ is nilpotent, there exists an m such that
$$(\mathrm{ad}_A)^{m+1}B = B^{(m+1)} = 0.$$

Set $N = m + \alpha_i$, $f(x) = p_i(x)^N$, $a = A$ and $b = B$. One has

$$f(A)B = Bf(A) + B'f'(A) + \cdots + B^{(m)}\frac{1}{m!}f^{(m)}(A).$$

Therefore, one has, for all $v \in V_i$

$$p_i(A)^N Bv = Bf(A)v + B'f'(A)v + \cdots = 0$$

and hence V_i is B-invariant. From here, Theorem $\tilde{3}$ follows readily. □

2.2. Solvable Lie algebras

2.2.1. Theorem of Lie

Let $\mathfrak{g} \subset \mathfrak{gl}(V)$, $\Bbbk = \mathbb{C}$, be a solvable Lie algebra. Then there exists a common eigenvector of $\mathfrak{g}$, namely
$$b \neq 0, \ Xb = \lambda(X)b, \ \forall X \in \mathfrak{g}.$$

Proof: The solvability assumption implies that
$$D\mathfrak{g} = [\mathfrak{g}, \mathfrak{g}] \subsetneq \mathfrak{g}, \ \mathfrak{g}/\mathfrak{D}\mathfrak{g} \text{ is abelian},$$

thus having a codimension one ideal $\mathfrak{g}_1 \supseteq D\mathfrak{g}$, which is also solvable. Let us prove the theorem by induction on $\dim \mathfrak{g}$. Applying the induction assumption to $\mathfrak{g}_1$, one has a common eigenvector $b \in V$ for $\mathfrak{g}_1$, namely,
$$Xb = \lambda(X)b, \ \forall X \in \mathfrak{g}_1.$$

Set W_λ to be the subspace of V satisfying

$$W_\lambda = \{v \in V; Xv = \lambda(X)v, \ \forall X \in \mathfrak{g}_1\}$$

and let $Y \notin \mathfrak{g}_1$ be a chosen element of $\mathfrak{g}$ outside of $\mathfrak{g}_1$. Moreover, for a chosen $v_0 \neq 0$ in W_λ, there exists an l such that

$$\{Y^j v_0; 0 \le j \le l-1\} \text{ is linearly independent, but}$$

$$\{Y^j v_0; 0 \le j \le l\} \text{ is linearly dependent.}$$

Set U to be the span, which is clearly Y-invariant. It is easy to check that, for $Z \in \mathfrak{g}_1$, one has

$$Z(Y^j v_0) \equiv \lambda(Z)Y^j v_0 \ (\mathrm{mod} \ v_0, \ldots, Y^{j-1}v_0),$$

thus proving that U is also $\mathfrak{g}_1$-invariant and

$$\mathrm{Tr}(Z|_U) = l\lambda(Z), \ \forall Z \in \mathfrak{g}_1.$$

In particular, one has

$$\mathrm{Tr}([Y,X]|_U) = l\lambda([Y,X]), \ \forall X \in \mathfrak{g}_1.$$

On the other hand, it follows from the Y and X-invariance property of U that $\mathrm{Tr}([Y,X]|_U) = 0$, thus having

$$\lambda([Y,X]) = 0, \ \forall X \in \mathfrak{g}_1$$

and hence

$$X(Yv) = Y(Xv) - [X,Y]v = \lambda(X)Yv, \ \forall v \in W_\lambda, X \in \mathfrak{g}_1$$

(i.e. W_λ is Y-invariant). Since $\Bbbk = \mathbb{C}$, there exists an eigenvector $\mathbf{b} \in W_\lambda$ of $Y|_{W_\lambda}$, which is a common eigenvector of $\mathfrak{g}$ that we are seeking.   □

Corollary 3.   *There exists a suitable basis of V so that $\mathfrak{g}$ consists of upper triangular matrices, i.e. of the type of Example 1.*

Corollary 4.   *Let $\mathfrak{g} \subset \mathfrak{gl}(V)$ be a solvable sub-Lie algebra, $\Bbbk = \mathbb{C}$. Then $D\mathfrak{g} = [\mathfrak{g}, \mathfrak{g}]$ is nilpotent.*

2.2.2. *Cartans criterion of solvability*

Theorem 4 (Cartan). *Let* $\mathfrak{g} \subset \mathfrak{gl}(V)$ *and* $\mathrm{char}(\Bbbk) = 0$. *Then* $\mathfrak{g}$ *is solvable if and only if*

$$\mathrm{Tr}(XY) = 0, \quad \forall X \in \mathfrak{g}, \quad Y \in D\mathfrak{g}.$$

Proof: First of all, it is easy to check that both solvability and the above trace condition are preserved under the extension of coefficient field $\Bbbk \subset \mathbb{C}$, thus the proof of Theorem 4 can be reduced to the special case of $\mathbb{C}$.

Secondly, if $\mathfrak{g}$ is solvable and $\Bbbk = \mathbb{C}$, it follows from Corollaries 3 and 4 of Theorem of Lie that $\mathfrak{g}$ (resp. $D\mathfrak{g}$) are conjugate to sub-Lie algebras of upper triangular (resp. upper triangular with zeros along the diagonal) matrices, and hence

$$\mathrm{Tr}(XY) = 0, \quad \forall X \in \mathfrak{g}, \quad Y \in D\mathfrak{g}.$$

Therefore it suffices to prove the sufficiency as follows:

Let us first state and prove the following lemma.

Lemma 2. *Let* $u \in \mathfrak{gl}(V)$, $\mathrm{ad}_u : \mathfrak{gl}(V) \to \mathfrak{gl}(V)$, $\mathrm{ad}_u(x) = [u, x]$. *Then the Jordan decompositions of* u *(resp.* ad_u*) satisfy the following relationship, namely*

$$s(\mathrm{ad}_u) = \mathrm{ad}_{s(u)}, n(\mathrm{ad}_u) = \mathrm{ad}_{n(u)}, \bar{s}(\mathrm{ad}_u) = \mathrm{ad}_{\bar{s}(u)}.$$

Proof: By Theorem 1, both of the components of u (resp. ad_u) are characterized by their properties. Therefore, it suffices to check that $\{\mathrm{ad}_{s(u)}, \mathrm{ad}_{n(u)}, \text{ and } \mathrm{ad}_{\bar{s}(u)}\}$ actually satisfy that of $\{s(\mathrm{ad}_u), n(\mathrm{ad}_u), \text{ and } \bar{s}(\mathrm{ad}_u)\}$.

It is easy to see that $\mathrm{ad}_{n(u)}$ is nilpotent by Engel's Theorem, and both $\mathrm{ad}_{s(u)}$ and $\mathrm{ad}_{\bar{s}(u)}$ are semi-simple with respective the basis of $e(\boldsymbol{b}_i, \boldsymbol{b}_j)$ of $\mathfrak{gl}(V)$ canonically associated with the eigenbasis of $s(u)$, namely, having

$$e(\boldsymbol{b}_i, \boldsymbol{b}_j) \cdot \boldsymbol{b}_k = \delta_{ik}\boldsymbol{b}_j,$$

$$\mathrm{ad}_{s(u)}(e(\boldsymbol{b}_i, \boldsymbol{b}_j)) = (\lambda_j - \lambda_i)e(\boldsymbol{b}_i, \boldsymbol{b}_j),$$

$$\mathrm{ad}_{\bar{s}(u)}(e(\boldsymbol{b}_i, \boldsymbol{b}_j)) = (\bar{\lambda}_j - \bar{\lambda}_i)e(\boldsymbol{b}_i, \boldsymbol{b}_j),$$

and, of course, also having the commutativity and

$$\mathrm{ad}_u = \mathrm{ad}_{s(u)} + \mathrm{ad}_{n(u)}.$$

Therefore, Lemma 2 follows from the uniqueness of Theorem 1. $\quad\square$

Proof: Now, by Lemmas 1 and 2, the proof of *sufficiency* amounts to prove that the above condition implies

$$\mathrm{Tr}(u\bar{s}(u)) = 0, \quad \forall u \in D\mathfrak{g} \text{ (i.e. } u = \sum c_i[x_i y_i]).$$

Using the well-known formula that

$$\mathrm{Tr}([A,B]C) = \mathrm{Tr}(B[C,A])$$

one has,

$$\mathrm{Tr}(u\bar{s}(u)) = \sum c_i \mathrm{Tr}(y_i[\bar{s}(u), x_i]) = \sum c_i \mathrm{Tr}(y_i \mathrm{ad}_{\bar{s}(u)}(x_i)).$$

By Lemma 2 and $\mathrm{ad}_u(\mathfrak{g}) \subset D\mathfrak{g}$, $\mathrm{ad}_{\bar{s}(u)} = \bar{s}(\mathrm{ad}_u)$ is a polynomial of ad_u *without* constant term, thus proving that

$$\mathrm{Tr}(u\bar{s}(u)) \;=\; 0, \quad \forall u \in D\mathfrak{g}$$

$$\Longrightarrow D\mathfrak{g} \text{ is nilpotent (by Lemma 1)}$$

$$\Longrightarrow \mathfrak{g} \text{ is solvable.} \qquad\qquad \square$$

3. Semi-Simple Lie Algebras

Recall that the radical of a Lie algebra $\mathfrak{g}$, denoted by $\mathfrak{r}$, is the unique maximal solvable ideal of $\mathfrak{g}$; $\mathfrak{g}/\mathfrak{r}$ is semi-simple and $\mathfrak{g}$ is an extension of $\mathfrak{g}/\mathfrak{r}$ by $\mathfrak{r}$, namely,

$$0 \to \mathfrak{r} \to \mathfrak{g} \to \mathfrak{g}/\mathfrak{r} \to 0.$$

In the case of $\mathrm{char}(\Bbbk) = 0$, there is the following criterion of semi-simplicity, corresponding to that of solvability.

Theorem 5 (Cartan). *A Lie algebra $\mathfrak{g}$ over a characteristic zero field is semi-simple if and only if the Killing bilinear form $B(X,Y) = \mathrm{Tr}(\mathrm{ad}_X \mathrm{ad}_Y)$ is non-degenerate.*

Proof: *Necessity*: Let $\mathfrak{g}$ be a semi-simple Lie algebra. Let $\mathfrak{a}$ be the following linear subspace, namely,

$$\mathfrak{a} = \{X \in \mathfrak{g}; B(X,Y) = 0, \forall Y \in \mathfrak{g}\}.$$

Note that the Killing bilinear form has the following basic invariance property:

$$B([Z,X],Y) + B(X,[Z,Y]) = 0.$$

It is easy to check that $\mathfrak{a}$ is an ideal of $\mathfrak{g}$, containing the center of $\mathfrak{g}$. Therefore, a direct application of Theorem 4 to $\mathrm{ad}_{\mathfrak{g}}|_{\mathfrak{a}} \subset \mathfrak{gl}(V)$ proves that $\mathrm{ad}_{\mathfrak{g}}|_{\mathfrak{a}}$ is solvable, and hence $\mathfrak{a}$ itself is also solvable.

Hence, the semi-simplicity of $\mathfrak{g}$ implies that $\mathfrak{a} = \{0\}$, i.e. $B(X,Y)$ is non-degenerate.

Sufficiency: Suppose $\mathfrak{g}$ is not semi-simple. Then its radical $\mathfrak{r}$ is non-trivial. Note that the last non-zero one of $\{D^n\mathfrak{r}\}$ is a commutative ideal, say denoted by $\mathfrak{a}$, of $\mathfrak{g}$. Let $X \in \mathfrak{a}, Y \in \mathfrak{g}$ and $A = \mathrm{ad}_X\mathrm{ad}_Y$. Then

$$A\mathfrak{g} \subset \mathfrak{a}, A\mathfrak{a} = 0, A^2 = 0$$

$$B(X,Y) = \mathrm{Tr}A = 0.$$

This proves that $B(-,-)$ is degenerate and also proves the sufficiency. $\square$

Corollary 5. *Suppose $\mathfrak{g}$ is semi-simple and $\mathfrak{a}$ is an ideal of $\mathfrak{g}$. Then*

$$\mathfrak{a}^\perp = \{X \in \mathfrak{g}; B(X,Y) = 0, \forall Y \in \mathfrak{a}\}$$

is also an ideal of $\mathfrak{g}$ and $\mathfrak{g} = \mathfrak{a} \oplus \mathfrak{a}^\perp$.

Proof: It follows readily from the non-degeneracy and invariance property of the Killing bilinear form. $\square$

Corollary 6. *A semi-simple Lie algebra $\mathfrak{g}$, $\mathrm{char}(\Bbbk) = 0$, has a direct sum decomposition*

$$\mathfrak{g} = \sum_{i=1}^{l} \mathfrak{g}_i$$

where the $\mathfrak{g}_i$, are simple Lie algebras (i.e. containing no non-trivial ideals).

Corollary 7. *Let $\mathfrak{g}$ be a semi-simple Lie algebra, $\mathrm{char}(\Bbbk) = 0$. Then $\mathfrak{g}\otimes_\Bbbk K$ is also semi-simple for any field $K \supset \Bbbk$.*

For example, the *complexification* of a compact semi-simple Lie algebra $\mathfrak{u}$, i.e. $\mathfrak{g} = \mathfrak{u} \otimes \mathbb{C}$, is of course, a semi-simple complex Lie algebra.

Lecture 8

Classification Theory of Complex Semi-Simple Lie Algebras

Historically, the classification theory of complex semi-simple Lie algebras is the monumental contribution of W. Killing, achieved at a very early stage of such studies. It has been traditionally the important core part of the theory of Lie groups and Lie algebras. In retrospect, the final result of his classification can be concisely restated as follows:

Theorem 1 (Killing). *A complex semi-simple Lie-algebra* $\mathfrak{g}$ *is always the complexification of a compact semi-simple Lie algebra, namely,*

$$\mathfrak{g} = \mathfrak{u} \otimes \mathbb{C}$$

and moreover,

$$\mathfrak{g}_1 \simeq \mathfrak{g}_2 \Longleftrightarrow \mathfrak{u}_1 \simeq \mathfrak{u}_2.$$

In this lecture, we shall present a proof of the above theorem with the benefits of hindsight and the knowledge of compact Lie algebras (cf. Lectures 5 and 6). In fact, the proof of the above theorem amounts to show that $\mathfrak{g}$ has a decomposition which is exactly the same as the complexified Cartan decomposition of $\mathfrak{u} \otimes \mathbb{C}$ (cf. Section 2, Lecture 5).

1. Cartan Subalgebras and Cartan Decompositions for the General Case of an Infinite Field $\Bbbk$

Let $\mathfrak{g}$ be a Lie algebra over $\Bbbk$, which has infinite number of elements. In the special case of a compact Lie algebra $\mathfrak{u}$, a Cartan subalgebra (resp. decomposition) is a maximal abelian subalgebra $\mathfrak{h}$ (resp. the primary decomposition of $\mathrm{ad}_\mathfrak{u}|_\mathfrak{h}$). Therefore, it is quite natural to seek their generalization in the setting of *nilpotent* subalgebras (resp. their primary decompositions), namely,

Definition A *nilpotent* subalgebra $\mathfrak{h}$ of $\mathfrak{g}$ is called a Cartan subalgebra of $\mathfrak{g}$ if

$$[X, \mathfrak{h}] \subseteq \mathfrak{h} \Longrightarrow X \in \mathfrak{h}.$$

Let $\{e_i; 1 \le i \le m = \dim \mathfrak{g}\}$ be a basis of $\mathfrak{g}$ and $X = \sum \xi_i e_i$ be a generic element of $\mathfrak{g}$. Set

$$f_X(\lambda) = \det(\mathrm{ad}_X - \lambda I) = (-1)^m[\lambda^m + g_1(\xi)\lambda^{m-1} + \cdots + g_{m-l}(\xi)\lambda^{m-l}]$$

where $g_i(\xi), 1 \le i \le m - l$, are homogeneous polynomials in $\xi = (\xi_1,\ldots,\xi_m)$, $g_{m-l}(\xi) \ne 0$ but $g_i(\xi) \equiv 0$ for $i > (m - l)$. Note that $\mathrm{ad}_X(X) \equiv 0$, thus having $l \ge 1$.

Definition An element $a \in \mathfrak{g}$ is called a *regular* element if $g_{m-l}(a) \ne 0$. Such regular elements exist because $|\Bbbk| = \infty$.

Lemma 1. *Let a be a regular element of $\mathfrak{g}$ and $\mathfrak{h}$ be the nilpotent subspace of ad_a, namely,*

$$\mathfrak{h} = \{X \in \mathfrak{g}; (\mathrm{ad}_a)^m X = 0\}.$$

Then $\mathfrak{h}$ is a Cartan subalgebra of $\mathfrak{g}$.

Proof: First of all, $\mathfrak{h}$ is a sub-Lie algebra. Suppose $X, Y \in \mathfrak{h}$. Then

$$(\mathrm{ad}_a)^m X = 0, (\mathrm{ad}_a)^m y = 0 \Longrightarrow (\mathrm{ad}_a)^{2m}[X, Y] = 0.$$

Secondly, we shall show that $\mathfrak{h}$ is nilpotent. Let $b \in \mathfrak{h}$ be an element of $\mathfrak{h}$; it suffices to show that $\mathrm{ad}_b|_\mathfrak{h}$ is nilpotent as follows (i.e. by Engel's Theorem).

Let $m_A(X)$ be the minimal polynomial of $A = \text{ad}_a$

$$m_A(X) = X^{\alpha_0} g(X), g(0) \neq 0$$

and

$$\mathfrak{g} = \mathfrak{h} \oplus V_1, V_1 = \ker g(A).$$

Let $B = \text{ad}_b$, $b \in \mathfrak{h}$. Then, $(\text{ad}_A)^{\alpha_0} B = 0$, thus having (cf. the proof of Theorem $\tilde{3}$)

$$g(A)^{\alpha_0+1} Bv = 0, \forall v \in V_1$$
$$\implies Bv = [b, v] \in \ker g(A)^{\alpha_0+1} = V_1.$$

Now, choose bases in $\mathfrak{h}$ and V_1, respectively, to have a combined basis of $\mathfrak{g}$. It follows from the $\{A, B\}$ invariance of both $\mathfrak{h}$ and V_1 that the matrices of A and B are of the following block form, namely

$$A \to \begin{pmatrix} (\alpha_1) & 0 \\ 0 & (\alpha_2) \end{pmatrix}, \quad B \to \begin{pmatrix} (\beta_1) & 0 \\ 0 & (\beta_2) \end{pmatrix}, \quad \det(\alpha_2) \neq 0.$$

We claim that (β_1) must be also nilpotent. For otherwise,

$$\det(\lambda I_m - c_1 A - c_2 B)$$
$$= \det(\lambda I_l - c_1(\alpha_1) - c_2(\beta_1)) \cdot \det(\lambda I_{m-l} - c_1(\alpha_2) - c_2(\beta_2))$$
$$= f_1(\lambda, c_1, c_2) \cdot f_2(\lambda, c_1, c_2)$$

while the factor of λ in $f_1(\lambda, 0, 1)$ is lower than l and $f_2(\lambda, 1, 0)$ has no λ factor whatsoever. Therefore, there exists (c_1, c_2) in $\Bbbk$ so that the characteristic polynomial of $c_1 A + c_2 B = \text{ad}_{(c_1 a + c_2 b)}$ does not have a factor of λ^l, contradicting the *minimality* of l in the definition of regular element.

Finally, let us verify that $[X, \mathfrak{h}] \subseteq \mathfrak{h}$ as follows:

$$\text{ad}_a(X) = [a, X] \in \mathfrak{h}$$
$$\implies (\text{ad}_a)^{m+1}(X) = 0 \Rightarrow X \in \mathfrak{h}. \qquad \square$$

We shall call the *primary* decomposition of $\text{ad}_\mathfrak{g}|_\mathfrak{h}$ a *Cartan decomposition* of $\mathfrak{g}$.

Remark. The above results hold in general as long as the coefficient field $\Bbbk$ contains infinitely many elements.

2. On the Structure of Cartan Decomposition of Complex Semi-Simple Lie Algebras

In this section we shall specialize to the specific case of complex semi-simple Lie algebras $\mathfrak{g}$. Let $\mathfrak{h}$ be a Cartan subalgebra of such a $\mathfrak{g}$ (cf. Lemma 1) and

$$\mathfrak{g} = \mathfrak{h} \oplus \sum_{\alpha \in \Delta} \mathfrak{g}_\alpha$$

where $\alpha \in \Delta$ are those non-zero complex linear functions of $\mathfrak{h}$ with

$$\mathfrak{g}_\alpha = \{X \in \mathfrak{g}; (\mathrm{ad}_H - \alpha(H)I_\mathfrak{g})^N X = 0, \quad \forall H \in \mathfrak{h}\} \neq \{0\}.$$

Naturally, the *non-degenerate* Killing bilinear form will play a major role in the discussion of this section, while our task is to show the above decomposition also has the same kind of properties as that of $\mathfrak{u} \otimes \mathbb{C}$ (cf. Section 2 of Lecture 5).

(i) $[\mathfrak{g}_\alpha, \mathfrak{g}_\beta] \subseteq \mathfrak{g}_{\alpha+\beta}, (\mathfrak{g}_{\alpha+\beta} = \mathfrak{h}$ in case $\alpha + \beta = 0$).

Proof: Let $H \in \mathfrak{h}, X_\alpha \in \mathfrak{g}_\alpha$ and $X_\beta \in \mathfrak{g}_\beta$, namely,

$$(\mathrm{ad}_H - \alpha(H)I_\mathfrak{g})^m X_\alpha = (\mathrm{ad}_H - \beta(H)I_\mathfrak{g})^m X_\beta = 0.$$

Note that

$$\mathrm{ad}_H([u, v]) = [\mathrm{ad}_H(u), v] + [u, \mathrm{ad}_H(v)]$$

and hence

$$(\mathrm{ad}_H - \alpha(H)I_\mathfrak{g} - \beta(H)I_\mathfrak{g})[u, v]$$
$$= [(\mathrm{ad}_H - \alpha(H)I_\mathfrak{g})(u), v] + [u, (\mathrm{ad}_H - \beta(H)I_\mathfrak{g})(v)].$$

Therefore, one has, by a kind of Leibniz formula

$$(\mathrm{ad}_H - \alpha(H)I_\mathfrak{g} - \beta(H)I_\mathfrak{g})^{2m}[X_\alpha, X_\beta]$$
$$= \sum_{i=0}^{2m} \binom{2m}{i} [(\mathrm{ad}_H - \alpha(H)I_\mathfrak{g})^{2m-i}X_\alpha, (\mathrm{ad}_H - \beta(H)I_\mathfrak{g})^i X_\beta] = 0$$

namely, $[\mathfrak{g}_\alpha, \mathfrak{g}_\beta] \subseteq \mathfrak{g}_{\alpha+\beta}$. $\qquad\square$

(ii) *Corollary of (i)* Suppose that $\alpha, \beta \in \Delta \cup \{0\}, \alpha + \beta \neq 0$. Then

$$B(X_\alpha, X_\beta) = 0, X_\alpha \in \mathfrak{g}_\alpha, X_\beta \in \mathfrak{g}_\beta.$$

Proof: For $Y_\gamma \in \mathfrak{g}_\gamma$ (resp. $\mathfrak{h}$ in case $\gamma = 0$), one has

$$(\mathrm{ad}_{X_\alpha} \mathrm{ad}_{X_\beta})Y_\gamma \in \mathfrak{g}_{\alpha+\beta+\gamma}.$$

Therefore, the diagonal elements of $\mathrm{ad}_{X_\alpha}\mathrm{ad}_{X_\beta}$ with respect to a basis of $\mathfrak{g}$ compatible with the Cartan decomposition are all zero, thus having

$$B(X_\alpha, X_\beta) = \mathrm{Tr}(\mathrm{ad}_{X_\alpha}\mathrm{ad}_{X_\beta}) = 0. \qquad \square$$

(iii) *Corollary of (ii)* The restriction of $B(-,-)$ onto $\mathfrak{h}$ is also non-degenerate; $\alpha \in \Delta \implies -\alpha \in \Delta$, and moreover, to an $X_\alpha \neq 0$ in $\mathfrak{g}_\alpha$, $\exists X_{-\alpha} \in \mathfrak{g}_{-\alpha}$ with $B(X_\alpha, X_{-\alpha}) \neq 0$.

Proof: Straightforward verification based on (ii) and the non-degeneracy of $B(-,-)$. $\qquad \square$

(iv) For $H_1, H_2 \in \mathfrak{h}$, one has

$$B(H_1, H_2) = \sum_{\alpha \in \Delta} \dim \mathfrak{g}_\alpha \alpha(H_1)\alpha(H_2).$$

Proof: Note that every $\mathfrak{g}_\alpha$ is ad_{H_i} invariant and matrices of $\mathrm{ad}_{H_i}|_{\mathfrak{g}_\alpha}$ are upper triangular with $\alpha(H)$ along the diagonal. Hence,

$$B(H_1, H_2) = \mathrm{Tr}(\mathrm{ad}_{H_1}\mathrm{ad}_{H_2}) = \sum_{\alpha \in \Delta} \dim \mathfrak{g}_\alpha \alpha(H_1)\alpha(H_2). \qquad \square$$

Lemma 2. *The Cartan subalgebras of a complex semi-simple Lie algebra $\mathfrak{g}$ is commutative.*

Proof: Let $H \in [\mathfrak{h}, \mathfrak{h}]$, namely,

$$H = \sum c_i[H_i, H_i'], H_i, H_i' \in \mathfrak{h}.$$

Then, it is easy to see that the matrices of $\mathrm{ad}_H|_{\mathfrak{g}_\alpha}$ are all upper triangular and with zero along their diagonals. Therefore,

$$B(H, H') = \mathrm{Tr}(\mathrm{ad}_H \mathrm{ad}_{H'}) = 0 \quad \forall H' \in \mathfrak{h}$$

and hence, by (iii), $H = 0$. This proves that $[\mathfrak{h}, \mathfrak{h}] = 0$. $\qquad \square$

Notation By (iii), there exists a unique $H_\alpha \in \mathfrak{h}$ such that

$$B(H_\alpha, H) = \alpha(H), \quad \forall H \in \mathfrak{h}.$$

(v) It follows from (iii) and (iv) that

$$\bigcap_{\alpha \in \Delta} \ker(\alpha) = \{0\}$$

and $\{H_\alpha, \alpha \in \Delta\}$ forms a generator system of $\mathfrak{h}$.

(vi) Let $X_\alpha \neq 0$ be a common eigenvector of $\mathrm{ad}_\mathfrak{h}$, namely,

$$[H, X_\alpha] = \alpha(X)X_\alpha, \quad \forall H \in \mathfrak{h}.$$

Then

$$B([X_\alpha, X_{-\alpha}], H) = B(X_{-\alpha}, [H, X_\alpha])$$
$$= B(X_{-\alpha}, \alpha(H)X_\alpha) = B(X_{-\alpha}, X_\alpha)B(H_\alpha, H).$$

Therefore,

$$B([X_\alpha, X_{-\alpha}] - B(X_\alpha, X_{-\alpha})H_\alpha, H) = 0, \quad \forall H \in \mathfrak{h},$$

and hence

$$[X_\alpha, X_{-\alpha}] = B(X_\alpha, X_{-\alpha})H_\alpha.$$

(We may choose $X_{-\alpha}$ such that $B(X_\alpha, X_{-\alpha}) = 1$.)

Lemma 3. *For each $\alpha \in \Delta$, one has $\alpha(H_\alpha) \neq 0$, dim $\mathfrak{g}_\alpha = 1$ and $j\alpha \in \Delta$ if and only if $j = \pm 1$.*

Proof: Let $X_\alpha \neq 0$ be a common eigenvector of $\mathrm{ad}_\mathfrak{h}$. Set

$$W = \mathbb{C}X_\alpha + \mathbb{C}H_\alpha + \sum_{j=1,2,\dots} \mathfrak{g}_{-j\alpha}.$$

By (vi), W is a sub-Lie algebra of $\mathfrak{g}$. Let us first prove $\alpha(H_\alpha) \neq 0$ by contradiction. Suppose the contrary that $\alpha(H_\alpha) = 0$. Then

$$W_0 = \mathbb{C}H_\alpha + \sum \mathfrak{g}_{-j\alpha} \subset W$$

would be a solvable ideal of W and hence W would also be solvable. Applying Theorem 7.2 to ad_W, there exists a basis of $\mathfrak{g}$ with respect to which the matrices of all $\{\mathrm{ad}_Y, Y \in W\}$ are upper triangular. Therefore the matrix of

$$\mathrm{ad}_{H_\alpha} = [\mathrm{ad}_{X_\alpha}, \mathrm{ad}_{X_{-\alpha}}]$$

is upper triangular with zero along the diagonal, namely, $\beta(H_\alpha) = 0$ $\forall \beta \in \Delta$, which contradicts the fact $H_\alpha \neq 0$ and (iii).

Note that W is invariant under both ad_{X_α} and $\mathrm{ad}_{X_{-\alpha}}$ and we may choose $X_{-\alpha}$ so that $H_\alpha = [X_\alpha, X_{-\alpha}]$. Therefore,

$$\mathrm{Tr}(\mathrm{ad}_{H_\alpha}|_W) = 0$$

on the one hand, and on the other hand

$$\mathrm{Tr}(\mathrm{ad}_{H_\alpha}|_W) = \alpha(H_\alpha) \cdot \{1 - \dim \mathfrak{g}_{-\alpha} - 2\dim \mathfrak{g}_{-2\alpha} - \cdots\}$$

where $\dim \mathfrak{g}_{-\alpha} \neq 0$. Hence, $\dim \mathfrak{g}_{-\alpha} = 1$ and moreover $\dim \mathfrak{g}_\alpha = 1$, and $\dim \mathfrak{g}_{-j\alpha} = 0$ for $j > 1$. $\qquad\square$

Summary At this junction, let us pause for a moment for a brief summary of the above algebraic structure of the specific features of Cartan decompositions of the special but also of basic importance case of complex, semi-simple Lie algebras.

(1) Lemmas 2 and 3 show that the Cartan decomposition of complex semi-simple Lie algebras also have the main features of the complex Cartan decomposition of compact semi-simple Lie algebras (cf. Section 2 of Lecture 5).

(2) Technically, such remarkable special features of the Cartan decomposition of $\mathfrak{u} \otimes \mathbb{C}$ are direct consequences of the maximal tori theorem and Schur's Lemma, while the proofs of Lemmas 1, 2 and 3 involve sequences of intricate applications of basic theorems on nilpotent, solvable and semi-simple Lie algebras, a "tour de force" of linear algebra on linear transformations.

(3) As a corollary of Lemma 3, for each $\alpha \in \Delta$

$$W_\alpha = \mathbb{C}X_\alpha \oplus \mathbb{C}H_\alpha \oplus \mathbb{C}X_{-\alpha}$$

is a sub-Lie algebra isomorphic to that of $SL(2,\mathbb{C})$, namely, the complexification of that of SU(2), having a basis $\{X'_\alpha, H'_\alpha, X'_{-\alpha}\}$ with

$$X'_\alpha = X_\alpha, H'_\alpha = \frac{2}{\alpha(H_\alpha)}H_\alpha, X'_{-\alpha} = \frac{2}{\alpha(H_\alpha)}X_{-\alpha}$$

satisfying

$$[X'_\alpha, X'_{-\alpha}] = H'_\alpha, [H'_\alpha, X'_\alpha] = 2X_\alpha, [H'_\alpha, X'_{-\alpha}] = 2X'_{-\alpha}.$$

Therefore, just in one stroke, it brings in the complex linear representation theory of SU(2) as a powerful tool in further detail analysis on the structure of the Cartan decomposition of $\mathfrak{g}$, just the *identical* way as in that of $\mathfrak{u} \otimes \mathbb{C}$. Anyhow, such an identical analysis also proves the existence of Chevalley basis (cf. Theorem 4 of Lecture 5) for $\mathfrak{g}$, thus giving a very precise explicit isomorphism of $\mathfrak{g} \simeq \mathfrak{u} \otimes \mathbb{C}$, namely, the Theorem of Killing reformulated.

Lecture 9

Lie Groups and Symmetric Spaces, the Classification of Real Semi-Simple Lie Algebras and Symmetric Spaces

Historically, in the development of the theory of Lie algebras, the great work of W. Killing on the classification of complex semi-simple Lie algebras was soon followed up by É. Cartan, on the classification theory of real semi-simple Lie algebras. On the other hand, in the study of Riemannian geometry, spaces of constant sectional curvature are outstanding important examples which are characterized by the symmetry property that the local isometry groups, $\text{ISO}(M^n, pt)$, are of the maximal possibility of $O(n)$ *everywhere*. In the case of simply connected ones, they are exactly those *reflectionally* symmetric spaces E^n, S^n, and H^n. A far-reaching generalization will be those spaces which are centrally symmetric with respect to every point $p \in M^n$, namely, $\text{ISO}(M^n, pt) \supset \{\pm \text{Id}\}$, nowadays simply referred to as *symmetric spaces*. Such a seemingly rather weak symmetry property *everywhere* turns out to be already extremely restrictive and can be effectively studied via Lie group theory. In fact, such symmetric spaces can be classified and the classification theory of symmetric spaces is intimately correlated with that of real semi-simple Lie algebras. This is the great contribution of É. Cartan, one of the most wonderful performances of geometric and algebraic theories developing hand in hand, almost like a beautiful musical duet.

We shall conclude our lectures on selected topics of Lie group theory by a concise presentation of some of the highlights of Cartan's contributions.

1. Real Semi-Simple Lie Algebras

Let $\mathfrak{g}$ be a real semi-simple Lie algebra and $\mathfrak{g}_{\mathbb{C}}$ be its *complexification*, i.e., $\mathfrak{g}_{\mathbb{C}} = \mathfrak{g} \oplus i\mathfrak{g}$, which is complex semi-simple Lie algebra that has already been classified by W. Killing. In particular, $\mathfrak{g}_{\mathbb{C}}$ also contains a compact semi-simple Lie algebra $\mathfrak{u}_0$ whose complexification is also $\mathfrak{g}_{\mathbb{C}}$, namely,

$$\mathfrak{g}_{\mathbb{C}} = \mathfrak{g} \oplus i\mathfrak{g} = \mathfrak{u}_0 \oplus i\mathfrak{u}_0$$

as Lie algebras over $\mathbb{R}$.

First of all, let us construct some examples of real semi-simple Lie algebras $\mathfrak{g}$ which are non-compact but have the same complexification as that of a given compact semi-simple Lie algebra $\mathfrak{u}_0$.

Example 1.   Let $\mathfrak{u}$ be a given compact semi-simple Lie algebra and σ be an involutive automorphism of $\mathfrak{u}$ (i.e. of order 2, or $\sigma^2 = \mathrm{Id}$). Then, one has

$$\mathfrak{u} = \ker(\sigma - \mathrm{Id}) \oplus \ker(\sigma + \mathrm{Id}) = \mathfrak{k} \oplus \mathfrak{p},$$
$$\mathfrak{k} = \{X \in \mathfrak{u}; \sigma(X) = X\}$$
$$\mathfrak{p} = \{X \in \mathfrak{u}; \sigma(X) = -X\}.$$

It is easy to check that

$$[\mathfrak{k}, \mathfrak{k}] \subseteq \mathfrak{k}, [\mathfrak{p}, \mathfrak{p}] \subseteq \mathfrak{k}, [\mathfrak{k}, \mathfrak{p}] \subseteq \mathfrak{p}$$

and hence $B(\mathfrak{k}, \mathfrak{p}) = 0$, while the Killing form $B(-, -)$ is negative definite because $\mathfrak{u}$ is compact semi-simple.

Set

$$\mathfrak{g} = \mathfrak{k} + i\mathfrak{p} = \{X + iY; X \in \mathfrak{k}, Y \in \mathfrak{p}\}.$$

It is easy to check that $\mathfrak{g} \subset \mathfrak{u} \otimes \mathbb{C}$ is a sub-Lie algebra over $\mathbb{R}$ and $\mathfrak{g} \otimes \mathbb{C} = \mathfrak{u} \otimes \mathbb{C}$, but it is a non-compact semi-simple Lie algebra. Note that

$$B(iY_1, iY_1) = -B(Y_1, Y_1) > 0 \; \forall Y_1 \neq 0 \text{ in } \mathfrak{p}.$$

The following theorem of É. Cartan proves that all real semi-simple Lie algebras can be obtained this way, namely,

Theorem 1 (Cartan's Lemma). *Let $\mathfrak{g}$ be a given real semi-simple Lie algebra and $\mathfrak{g}_{\mathbb{C}}$ be its complexification, that is,*

$$\mathfrak{g}_{\mathbb{C}} = \mathfrak{g} \otimes \mathbb{C} = \mathfrak{g} \oplus i\mathfrak{g} \; (over \; \mathbb{R}),$$

and let $\sigma : \mathfrak{g}_{\mathbb{C}} \to \mathfrak{g}_{\mathbb{C}}$ be the conjugation with respect to $\mathfrak{g}$. Then there exists a compact semi-simple subalgebra $\mathfrak{u}$ (of the real Lie algebra structure of $\mathfrak{g}_{\mathbb{C}}$) such that $\mathfrak{u} \otimes \mathbb{C} = \mathfrak{g}_{\mathbb{C}}$ and $\mathfrak{u}$ is invariant under σ.

Proof: Let σ (resp. τ_0) be the conjugation of $\mathfrak{g}_{\mathbb{C}}$ with respect to $\mathfrak{g}$ (resp. $\mathfrak{u}_0$), which are only $\mathbb{R}$-linear by themselves, but $\rho = \sigma\tau_0$ is an automorphism of the complex Lie algebra $\mathfrak{g}_{\mathbb{C}}$. However, $\mathfrak{u}_0$ is, in general, not invariant under σ. Therefore, the proof of Theorem 1 amounts to produce an automorphism φ of $\mathfrak{g}_{\mathbb{C}}$ such that $\varphi(\mathfrak{u}_0)$ is invariant under σ.

Suppose that φ is such a complex automorphism of $\mathfrak{g}_{\mathbb{C}}$. Then $\tau = \varphi\tau_0\varphi^{-1}$ is a conjugation of $\mathfrak{g}_{\mathbb{C}}$ with the fixed point set of τ, say denoted by $F(\tau)$, equal to $\varphi(\mathfrak{u}_0)$, namely,

$$F(\varphi\tau_0\varphi^{-1}) = \varphi F(\tau_0) = \varphi(\mathfrak{u}_0).$$

Therefore, $\mathfrak{u} = \varphi(\mathfrak{u}_0)$ is σ-invariant if $\sigma\tau = \tau\sigma$, namely,

$$\sigma F(\tau) = \sigma\tau F(\tau) = F(\sigma\tau\tau\sigma) = F(\tau).$$

Set $\rho = \sigma\tau_0$, which is a complex automorphism of $\mathfrak{g}_{\mathbb{C}}$, and set

$$\langle X, Y \rangle = -B(X, Y), X, Y \in \mathfrak{u}_0,$$

which is an inner product on $\mathfrak{u}_0$. Therefore,

$$\langle X, Y \rangle = -B(X, \tau_0 Y), X, Y \in \mathfrak{g}_{\mathbb{C}}$$

is a Hermitian inner product on $\mathfrak{g}_{\mathbb{C}}$. It follows from the invariant property of $B(-,-)$

$$B(\rho X, \tau_0 Y) = B(X, \rho^{-1}\tau_0 Y) = B(X, \tau_0\sigma\tau_0 Y) = B(X, \tau_0\rho Y),$$

namely, ρ is a Hermitian linear transformation of $\mathfrak{g}_{\mathbb{C}}$ and hence $P = \rho^2$ is a positive definite Hermitian linear transformation of $\mathfrak{g}_{\mathbb{C}}$. Therefore, there exists a basis of $\mathfrak{g}_{\mathbb{C}}$ consisting of eigenvectors of ρ (resp. P), namely,

$$\rho X_i = \mu_i X_i, \mu_i \in \mathbb{R}; PX_i = \lambda_i X_i, \lambda_i = \mu_i^2 > 0$$

for $1 \le i \le m = \dim \mathfrak{g}_{\mathbb{C}} = \dim_{\mathbb{R}} \mathfrak{g}$.

Set $\{C_{ij}^k\}$ to be the set of structure constants of $\mathfrak{g}_{\mathbb{C}}$ with respect to such a basis $\{X_i, 1 \le 1 \le m\}$, namely,

$$[X_i, X_j] = \sum_{k=1}^{m} C_{ij}^k X_k, \quad 1 \le i, j \le m.$$

Since P is an automorphism of $\mathfrak{g}_{\mathbb{C}}$, we have

$$\sum_{k=1}^{m} \lambda_i \lambda_j C_{ij}^k X_k = [\lambda_i X_i, \lambda_j X_j] = [PX_i, PX_j]$$

$$= P[X_i, X_j] = P\left(\sum_{k=1}^{m} C_{ij}^k X_k\right) = \sum_{k=1}^{m} \lambda_k C_{ij}^k X_k,$$

namely,

$$\lambda_i \lambda_j C_{ij}^k = \lambda_k C_{ij}^k, \quad 1 \le i, j, k \le m,$$

and hence

$$\lambda_i^t \lambda_j^t C_{ij}^k = \lambda_k^t C_{ij}^k, \quad \forall t \in \mathbb{R},$$

thus having $P^t, \forall t \in \mathbb{R}$, are complex automorphisms of $\mathfrak{g}_{\mathbb{C}}$ and moreover, it is easy to check that

$$\tau_0 P \tau_0 = P^{-1}, \quad \tau_0 P^t \tau_0 = P^{-t}.$$

Set $\varphi = P^{1/4}, \tau = \varphi \tau_0 \varphi^{-1}$. One has

$$\sigma\tau = \sigma P^{1/4} \tau_0 P^{-1/4} = \sigma\tau_0 P^{-1/2} = \rho P^{-1/2}$$
$$\tau\sigma = (\sigma\tau)^{-1} = P^{1/2}\rho^{-1}$$

and it is straightforward to check that

$$\rho P^{-1/2} X_i = \frac{\mu_i}{\sqrt{\lambda_i}} X_i = \text{sign}(\mu_i) X_i$$

$$P^{1/2} \rho^{-1} X_i = \frac{\sqrt{\lambda_i}}{\mu_i} X_i = \text{sign}(\mu_i) X_i$$

thus having $\sigma\tau = \tau\sigma, \mathfrak{u} = P^{1/4}(\mathfrak{u}_0)$ is σ-invariant. $\qquad\Box$

Corollary. *To a given real semi-simple Lie algebra $\mathfrak{g}$, there exists a unique compact semi-simple Lie algebra $\mathfrak{u}$ and an involutive automorphism σ such that*

$$\mathfrak{u} = \mathfrak{k} \oplus \mathfrak{p}, \mathfrak{g} = \mathfrak{k} \oplus i\mathfrak{p},$$

where

$$\mathfrak{k} = \ker(\sigma - \text{Id}), \mathfrak{p} = \ker(\sigma + \text{Id}).$$

2. Lie Groups and Symmetric Spaces

Definition 1. A Riemannian manifold M^n is called a *symmetric space* if it is *central symmetric* with respect to *every point* $p \in M^n$, namely, the local isometry group $\text{ISO}(M^n, p) \supset \{\pm\text{Id}\}, \forall p \in M^n$.

Note that those classical spaces of constant curvatures such as E^n, S^n, and H^n are characterized by the symmetry property of having $\text{ISO}(M^n, p) \simeq O(n), \forall p \in M^n$. Conceptually, the *ultimate weakening* of the symmetry condition from $\text{ISO}(M^n, p) \simeq O(n)$ to that of $\text{ISO}(M^n, p) \supseteq \{\pm\text{Id}\}$ for the definition of symmetric spaces presents a program of very far-reaching generalization of those *reflectionally* symmetric spaces. At first glance, it is almost unbelievable that such a bold program will naturally lead to a clean list of classification, together with that of real semi-simple Lie algebras.

Let M be a symmetric space, $G(M)$ be the isometry group of M which is a *differentiable* transformation group of M, and this will, of course, be our embarkation point of the journey of studying the structure of symmetric spaces.

Theorem 2. *Let M be a symmetric space. Then*

(i) *M is a complete Riemannian manifold.*

(ii) $G(M)$ *acts transitively on* M, *i.e.*, M *is a homogeneous space of the Lie group* $G(M)$.

(iii) *The isotropy subgroup* $K = G_{p_0}$ *is a compact Lie group;* $K = \mathrm{ISO}(M, p_0) \subseteq O(T_{p_0}M) \simeq O(n)$ *is the orthogonal isotropy representation of* G/K.

(iv) *To a given geodesic* $\gamma(t)$ *of* M, *there exists a unique one-parameter subgroup* $\varphi_\gamma : \mathbb{R} \to G(M)$ *such that*

$$\varphi_\gamma(t_1)\gamma(t_2) = \gamma(t_1 + t_2)$$
$$d\varphi_\gamma(t_1)|_{T_{\gamma(t_2)}M} = \|_\gamma(t_2, t_1 + t_2).$$

Proof: (i) Let us first prove the completeness of M. It suffices to show that any geodesic curve, γ, of M is infinitely prolongable [H-R]. Let $\gamma : [a, b] \to M$ be a geodesic parametrized by arc-length, and choose $\varepsilon < \frac{1}{2}(b - a)$. Set $\mathfrak{s}$ to be the central symmetry of M at $\gamma(b - \varepsilon)$. Then γ and $\mathfrak{s}(\gamma)$ has a matching segment of length 2ε and extending a segment of $(b - a - 2\varepsilon)$ beyond $\gamma(b)$.

(ii) Let $\{p, q\}$ be any given pair of points on M. By the completeness, there exists a shortest geodesic curve γ with $\gamma(0) = p$ and $\gamma(d(p, q)) = q$. Set $\mathfrak{s}$ to be the central symmetry of M at $\gamma(\frac{1}{2}d(p, q))$. Then $\mathfrak{s}(p) = q, \mathfrak{s}(q) = p$, thus proving the transitivity of $G(M)$ on M.

(iii) Let $K = \mathrm{ISO}(M, p_0), k \in K$. Then $dk|_{p_0} \in O(T_{p_0}M)$, and it follows from the unique existence of geodesics with given initial point and velocity, k is uniquely determined by $dk|_{p_0}$. Hence, the map $K \to O(n), k \mapsto dk|_{p_0}$ is injective, and it is exactly the isotropy representation of G/K.

(iv) Let $\gamma : \mathbb{R} \to M$ be a given geodesic curve parametrized by arc-length and $\mathfrak{s}_t$ be the central symmetry of M at $\gamma(t)$. Set $X(t)$ to be a parallel vector field defined on the segment $\gamma([t_1, t_2])$. Then

$$\mathfrak{s}_{t_1}X(t) = X'(2t_1 - t), \quad X'(t_1) = -X(t_1)$$

is a parallel vector field defined on $\gamma([2t_1 - t_2, t_1])$. Set

$$\varphi_\gamma(t_1) = \mathfrak{s}_{t_1/2}\mathfrak{s}_0.$$

It is easy to see that

$$\varphi_\gamma(t_1)\gamma(t_2) = \gamma(t_1 + t_2)$$
$$d\varphi_\gamma(t_1)|_{T_{\gamma(t_2)}M} = \|_\gamma(t_2, t_1 + t_2)$$

where $\|_\gamma$ is the parallel translation along γ. Therefore, it follows that

$$\varphi_\gamma(t_1)\varphi_\gamma(t_1') = \varphi_\gamma(t_1 + t_1'),$$
$$\varphi_\gamma \ : \ \mathbb{R} \to G(M)$$

forms a one-parameter subgroup of $G(M)$. Such a one-parameter subgroup of isometries, if it exists, is often referred to as the *transvections* along the geodesic γ. □

Corollary. *To any* $g \in G(M)$, *if* $g(p_0) \neq p_0$ (*i.e.* g *not in* K), *then there exists a geodesic* γ *with* $\gamma(d(p_0, g(p_0))) = g(p_0))$ *and hence,* $g\varphi_\gamma(-d) \in K$.

Theorem 3. *Let* M *be a symmetric space,* $G(M)$ *be its isometry group. For a chosen base point* p_0, *set* $\mathfrak{s}_0$ *to be central symmetry of* M *with* p_0 *as its center and set* $K = G_{p_0}$. *Then*

(i) $M \cong G(M)/K, K$ *is a compact Lie group.*

(ii) $\sigma : G(M) \to G(M), \sigma(g) = \mathfrak{s}_0 g \mathfrak{s}_0, \forall g \in G(M)$ *is an involutive automorphism of* $G(M)$.

(iii) *Set* $F(\sigma)$ (*resp.* $F^0(\sigma)$) *to be the set of fixed elements of* σ *in* $G(M)$ (*resp. the identity component of* $F(\sigma)$). *Then*

$$F^0(\sigma) \subseteq K \subseteq F(\sigma).$$

(iv) *Let* $\mathfrak{g}$ (*resp.* $\mathfrak{k}$) *be the Lie algebra of* $G(M)$ (*resp.* K). *Then, one has*

$$\mathfrak{g} = \mathfrak{k} \oplus \mathfrak{p}, \quad \mathfrak{k} = \ker(d\sigma - \mathrm{Id}), \quad \mathfrak{p} = \ker(d\sigma + \mathrm{Id}).$$

(v) *To any given* $X \in \mathfrak{p}, \gamma(t) = \mathrm{Exp}\,(tX(p_0))$ *is exactly the one-parameter subgroup of transvections along the unique geodesic with* p_0 (*resp.* $X(p_0)$) *as the initial position* (*resp. velocity*).

Proof: (i) By Theorem 2, $\iota : K \to O(T_{p_0}M) = O(n)$ is an injective isomorphism of K to a closed subgroup of $O(n)$ and hence K is a compact Lie group, $M = G(M)/K$ is a homogeneous Riemannian manifold.

(ii) Note that $\iota(s_0) = -\mathrm{Id} \in O(n)$. Therefore, one has

$$\iota(s_0 k s_0) = \iota(k), \text{i.e. } \sigma(k) = k, \forall k \in K, K \subseteq F(\sigma).$$

On the other hand, suppose that $\operatorname{Exp} tX$ is a one-parameter subgroup of $G(M)$ satisfying

$$s_0(\operatorname{Exp} tX)s_0 = \operatorname{Exp} tX, \text{ i.e. } \operatorname{Exp} tX \cdot s_0 \cdot \operatorname{Exp} - tX = s_0.$$

Then

$$\begin{aligned} s_0(\operatorname{Exp} tX \cdot p_0) &= \operatorname{Exp} tX s_0 \operatorname{Exp} - tX(\operatorname{Exp} tX p_0) \\ &= \operatorname{Exp} tX p_0 \quad \forall t \in \mathbb{R}, \end{aligned}$$

namely, $\operatorname{Exp} tX p_0$ is a connected subset of the fixed point set of s_0 containing p_0. However, p_0 is clearly an isolated point of the fixed point set of s_0. Hence

$$\operatorname{Exp} tX \in K, \quad K \supseteq F^0(\sigma, G(M)).$$

(iii) $d\sigma : \mathfrak{g} \to \mathfrak{g}$ is an involutive automorphism of $\mathfrak{g}$. Set

$$\mathfrak{k} = \ker(d\sigma - \operatorname{Id}), \quad \mathfrak{p} = \ker(d\sigma + \operatorname{Id})$$
$$\mathfrak{g} = \mathfrak{k} \oplus \mathfrak{p}.$$

By (iii), $\mathfrak{k}$ is the Lie algebra of K while

$$X \in \mathfrak{p} \Longleftrightarrow s_0 \operatorname{Exp} tX s_0 = \operatorname{Exp} - tX, \quad \forall t \in \mathbb{R}.$$

(iv) Let X_0 be the tangent vector of $\operatorname{Exp} tX_0 \cdot p_0$ at p_0 and γ be the unique geodesic curve with the initial point p_0 and X_0 as its initial velocity. Then, one has

$$\varphi_\gamma(t) = s_{t/2} \cdot s_0$$
$$s_0(\varphi_\gamma(t))s_0 = s_0 s_{t/2} \cdot s_0 s_0 = s_0 s_{t/2} = \varphi_\gamma(-t)$$

thus having

$$\operatorname{Exp} tX_0 = \varphi_\gamma(t), \quad \gamma(t) = \operatorname{Exp} tX_0(p_0). \qquad \square$$

3. Orthogonal Involutive Lie Algebras

Theorems 2 and 3 of Sec. 2 naturally lead to the study of the following type of Lie algebra structure, namely,

Definition. A real Lie algebra $\mathfrak{g}$ together with an involutive automorphism σ, and moreover, the decomposition

$$\mathfrak{g} = \mathfrak{k} \oplus \mathfrak{p};\, \mathfrak{k} = \ker(\sigma - \mathrm{Id}),\, \mathfrak{p} = \ker(\sigma + \mathrm{Id})$$

has the following properties, namely,

$$\mathrm{ad}_{\mathfrak{p}} : \mathfrak{k} \to \mathfrak{gl}(\mathfrak{p}), \quad \mathrm{ad}_{\mathfrak{p}}(X)Y = [X, Y]$$

is injective and $\mathfrak{p}$ is equipped with an $\mathrm{ad}_{\mathfrak{p}}\mathfrak{k}$ invariant inner product Q, i.e.

$$Q([X, Y], Z) + Q(Y, [X, Z]) \equiv 0, \quad \forall X \in \mathfrak{k} \quad \text{and} \quad Y, Z \in \mathfrak{p}.$$

Such a triple $(\mathfrak{g}, \sigma, Q)$ will be, henceforth, referred to as an *orthogonal involutive Lie algebra*.

Lemma 1. *Let $C(\mathfrak{g})$ be the center of $\mathfrak{g}$. Then*

$$C(\mathfrak{g}) \cap \mathfrak{k} = \{0\}$$

and the restriction of the Killing form of $\mathfrak{g}$ to $\mathfrak{k}$ is non-degenerate.

Proof: Let X (resp. Y) be elements of $C(\mathfrak{g}) \cap \mathfrak{k}$ (resp. $\mathfrak{p}$). Then $\mathrm{ad}_{\mathfrak{p}}(X) \cdot Y = [X, Y] = 0$ and by the injective assumption $X = 0$. Moreover, $\mathfrak{k}$ is a compact Lie algebra and $\mathfrak{p}$ is equipped with an $\mathrm{ad}_{\mathfrak{p}}\mathfrak{k}$-invariant inner product. Therefore,

$$B(X, X) \leq 0 \quad \forall X \in \mathfrak{k}$$

and equality holds when and only when $\mathrm{ad}X = 0$, thus X is equal to 0, namely, $B|\mathfrak{k}$ is negative definite. $\qquad\Box$

3.1. *Examples of orthogonal involutive Lie algebras*

Example 2 (cf. Example 1 of Sec. 1). Let $\mathfrak{u}$ be a compact simple Lie algebra and σ be an involutive automorphism of $\mathfrak{u}$.

$$\mathfrak{u} = \mathfrak{k} \oplus \mathfrak{p};\, \mathfrak{k} = \ker(\sigma - \mathrm{Id}),\, \mathfrak{p} = \ker(\sigma + \mathrm{Id}).$$

Set $-B(.,.)$ to be the inner product on $\mathfrak{u}$ and Q to be its restriction on $\mathfrak{p}$. Then $(\mathfrak{u}, \sigma, Q)$ is an orthogonal involutive Lie algebra.

Example 3. $\mathfrak{g} = \mathfrak{k} \oplus i\mathfrak{p}$, where $\mathfrak{k}$ and $\mathfrak{p}$ are the same as in Example 1. It is easy to see that $(\mathfrak{g}, \sigma, Q)$ with $Q(Y_1, Y_2) = B(Y_1, Y_2)$ is again an orthogonal involutive Lie algebra.

Example 4. Let M be a symmetric space and $\mathfrak{g}$ be the Lie algebra of $G(M)$. By Theorems 2 and 3, one has

$$\mathfrak{g} = \mathfrak{k} \oplus \mathfrak{p}, \quad \mathfrak{p} \cong T_{p_0}M \quad \text{(as inner product spaces)}.$$

Therefore $(\mathfrak{g}, \sigma, Q)$ is an orthogonal involutive Lie algebra, where Q is just the inner product of $T_{p_0}M$ and σ is the differential of that of $G(M)$ (cf. Theorem 3).

Example 4₀. In the very special case of the Euclidean space $\mathbf{E}^n$,

$$G(M) \cong O(n) \ltimes T^n$$
$$\mathfrak{g} = \mathfrak{o}(n) \oplus \mathbb{R}^n = \mathfrak{k} \oplus \mathfrak{p}$$

where T^n is the subgroup of translations, thus having $[\mathfrak{p}, \mathfrak{p}] = 0$. Furthermore, it is easy to check that direct sums of orthogonal involutive Lie algebras are again orthogonal involutive Lie algebras.

3.2. *On the direct sum decomposition of orthogonal involutive Lie algebra*

The additional (σ, Q)-structure of an orthogonal involutive Lie algebra $(\mathfrak{g}, \sigma, Q)$ makes the following direct sum decomposition readily available:

(1) Let us start with the Q-inner product structure on $\mathfrak{p}$. Note that $B(\cdot, \cdot)$ is invariant and $\mathrm{ad}_{\mathfrak{p}}K^0$ is Q-orthogonal. Therefore, there exists an orthonormal basis, $\{X_i, 1 \leq i \leq \dim \mathfrak{p} = n\}$, of $\mathfrak{p}$ with respect to Q such that

$$B\left(\sum_{i=1}^n \xi_i X_i, \sum_{i=1}^n \xi_i X_i\right) = \sum_{i=1}^n \lambda_i \xi_i^2$$
$$B\left(\sum_{i=1}^n \xi_i X_i, \sum_{i=1}^n \eta_i X_i\right) = \sum_{i=1}^n \lambda_i \xi_i \eta_i. \tag{1}$$

Set

$$\begin{cases} \mathfrak{p}_0 = \text{the span of } X_i & \text{with } \lambda_i = 0 \\ \mathfrak{p}_+ = \text{the span of } X_i & \text{with } \lambda_i > 0 \\ \mathfrak{p}_- = \text{the span of } X_i & \text{with } \lambda_i < 0. \end{cases}$$

Then

$$\mathfrak{p} = \mathfrak{p}_0 \oplus \mathfrak{p}_+ \oplus \mathfrak{p}_-,$$

where $\{\mathfrak{p}_0, \mathfrak{p}_+, \mathfrak{p}_-\}$ are $\mathrm{ad}_\mathfrak{p} \mathfrak{k}$ invariant and

$$B(\mathfrak{p}_0, \mathfrak{p}) = 0, \quad B(\mathfrak{p}_+, \mathfrak{p}_-) = 0 \quad \text{and} \quad B(\mathfrak{k}, \mathfrak{p}) = 0. \qquad (2)$$

Therefore

$$\mathfrak{p}_0 = \{X \in \mathfrak{g}; \ B(X, \mathfrak{p}) = 0\}$$

is an ideal of $\mathfrak{g}$ and moreover, commutative because

$$[\mathfrak{p}_0, \mathfrak{p}] \subset \mathfrak{k} \quad \text{and} \quad B(\mathfrak{k}, [\mathfrak{p}_0, \mathfrak{p}_0]) = 0$$
$$\Rightarrow [\mathfrak{p}_0, \mathfrak{p}_0] = \{0\}.$$

(2) Set

$$\mathfrak{k}_+ = [\mathfrak{p}_+, \mathfrak{p}_+], \quad \mathfrak{k}_- = [\mathfrak{p}_-, \mathfrak{p}_-]$$

and $\mathfrak{k}_0$ to be the orthogonal complement of $\mathfrak{k}_+ + \mathfrak{k}_-$ in $\mathfrak{k}$. Then, the following straightforward checking will show that $\{\mathfrak{k}_+, \mathfrak{k}_-, \mathfrak{k}_0\}$ are ideals of $\mathfrak{k}$ and

$$\mathfrak{k} = \mathfrak{k}_+ \oplus \mathfrak{k}_- \oplus \mathfrak{k}_0,$$

namely,

(i) $[\mathfrak{p}_+, \mathfrak{p}_-] \subset \mathfrak{k}$ and $B(\mathfrak{k}, [\mathfrak{p}_+, \mathfrak{p}_-]) = 0 \Rightarrow [\mathfrak{p}_+, \mathfrak{p}_-] = 0$.

(ii) $[\mathfrak{k}, \mathfrak{p}_\pm] \subset \mathfrak{p}_\pm$ and

$$[\mathfrak{k}, \mathfrak{k}_\pm] = [\mathfrak{k}, [\mathfrak{p}_\pm, \mathfrak{p}_\pm]] \subseteq [[\mathfrak{k}, \mathfrak{p}_\pm], \mathfrak{p}_\pm] \subseteq \mathfrak{k}_\pm.$$

Theorem 4. *An orthogonal involutive Lie algebra* $(\mathfrak{g}, \sigma, Q)$ *has the following direct sum decomposition, namely,*

$$(\mathfrak{g}, \sigma, Q) = (\mathfrak{g}_0, \sigma_0, Q_0) \oplus (\mathfrak{g}_+, \sigma_+, Q_+) \oplus (\mathfrak{g}_-, \sigma_-, Q_-)$$

where

$$\mathfrak{g}_0 = \mathfrak{k}_0 \oplus \mathfrak{p}_0, \mathfrak{g}_+ = \mathfrak{k}_+ \oplus \mathfrak{p}_+, \mathfrak{g}_- = \mathfrak{k}_- \oplus \mathfrak{p}_-$$

$$\sigma_0 = \sigma|\mathfrak{g}_0, \sigma_+ = \sigma|\mathfrak{g}_+, \sigma_- = \sigma|\mathfrak{g}_-$$

$$Q_0 = Q|\mathfrak{p}_0, Q_+ = Q|\mathfrak{p}_+, Q_- = Q|\mathfrak{p}_-$$

and moreover, $B|\mathfrak{p}_+$ *(resp.* $B|\mathfrak{p}_-$*) are positive (resp. negative) definite.*

Proof: It suffices to check that

$$[\mathfrak{k}_0, \mathfrak{p}_\pm] = \{0\}, [\mathfrak{k}_\pm, \mathfrak{p}_0] = \{0\}, [\mathfrak{k}_\pm, \mathfrak{p}_\mp] = \{0\}$$

as follows, namely,

(i) $[\mathfrak{k}_0, \mathfrak{p}_\pm] \subseteq \mathfrak{p}_\pm$ and $B([\mathfrak{k}_0, \mathfrak{p}_\pm], \mathfrak{p}_\pm) = 0 \Rightarrow [\mathfrak{k}_0, \mathfrak{p}_\pm] = 0$.

(ii) $[\mathfrak{k}_\pm, \mathfrak{p}_0] = [[\mathfrak{p}_\pm, \mathfrak{p}_\pm], \mathfrak{p}_0] \subseteq [\mathfrak{p}_\pm, [\mathfrak{p}_\pm, \mathfrak{p}_0]] = \{0\}$.

(iii) $[\mathfrak{k}_\pm, \mathfrak{p}_\mp] = [[\mathfrak{p}_\pm, \mathfrak{p}_\pm], \mathfrak{p}_\mp] \subset [\mathfrak{p}_\pm, [\mathfrak{p}_\pm, \mathfrak{p}_\mp]] = \{0\}$.   □

Corollary 1. *Let* $(\mathfrak{g}, \sigma, Q)$ *be an orthogonal involutive Lie algebra. Then* $\mathfrak{g}$ *is compact semi-simple if and only if* $\mathfrak{g} = \mathfrak{g}_-$.

Corollary 2. *Let* $(\mathfrak{g}, \sigma, Q)$ *be an orthogonal involutive Lie algebra and* $\mathfrak{g} = \mathfrak{g}_+$. *Then* $(\mathfrak{g}^*, \sigma^*, Q^*)$ *with*

$$\mathfrak{g}^* = \mathfrak{k} \oplus i\mathfrak{p}, \quad \sigma^* = \mathrm{Id}_\mathfrak{k} \oplus (-\mathrm{Id}_\mathfrak{p}),$$
$$Q^*(iX, iY) = Q(X, Y) \quad \forall X, Y \in \mathfrak{p}$$

is a compact semi-simple orthogonal involutive Lie algebra.

Corollary 3. *An orthogonal involutive Lie algebra* $(\mathfrak{g}, \sigma, Q)$ *can be decomposed into the direct sum* $(\mathfrak{g}_0, \sigma_0, Q_0)$ *and those irreducible components of* $(\mathfrak{g}_+, \sigma_+, \mathfrak{p}_+)$ *(resp.* $(\mathfrak{g}_-, \sigma_-, Q_-)$*).*

4. Classification Theory of Symmetric Spaces and Real Semi-Simple Lie Algebras

4.1. *A brief summary and overview*

Before we proceed to the final stage of classification theory of É. Cartan on symmetric spaces and real semi-simple Lie algebras, let us begin with a brief summary and overview of the structure theory on real semi-simple Lie algebras, symmetric spaces and orthogonal involutive Lie algebras (cf. Secs. 1–3).

(1) The structure of an orthogonal involutive Lie algebra naturally emerges from that of the isometry group of a given symmetric space M (cf. Theorems 2 and 3 and Example 3 of Sec. 3.1). Conceptually, it is just the linearization of $\{G(M),$ the conjugation of s_0 and the inner product Q on $T_{p_0}M\}$, almost the same way as that of the Lie algebra structure of a given Lie group. Moreover, it is easy to see the universal covering manifold of a symmetric space M, say denoted by $\tilde{M}$, equipped by the induced metric is again a symmetric space and having the same orthogonal involutive Lie algebra. Thus, the correspondence between simply connected symmetric spaces and their associated orthogonal involutive Lie algebras becomes

one-to-one. [Note that the isometry subgroup of the Euclidean factor of $\tilde{M}$, if any, should be that of Example 4_0.]

Anyhow, the classification of symmetric spaces can simply be reduced to that of semi-simple orthogonal involutive Lie algebras, which can be further reduced to that of irreducible ones (cf. Theorem 4).

(2) By Corollaries 1 and 2, one has a duality between irreducible orthogonal involutive Lie algebras of compact type (i.e. with $\mathfrak{g} = \mathfrak{g}_-$) and of non-compact type (i.e. with $\mathfrak{g} = \mathfrak{g}_+$). Thus, it suffices to consider either one of the above two types. Let us just consider the case of non-compact ones, namely, $\mathfrak{g}$ is a simple real Lie algebra of non-compact type.

(3) Let $\mathfrak{g}$ be a simple real Lie algebra of non-compact type and $\mathfrak{g}_{\mathbb{C}} = \mathfrak{g} \otimes \mathbb{C}$ be its complexification. Then, $\mathfrak{g}_{\mathbb{C}}$ is either a simple complex Lie algebra or

$$\mathfrak{g}_{\mathbb{C}} \cong \mathfrak{u}_1 \otimes \mathbb{C} \oplus \mathfrak{u}_1 \otimes \mathbb{C}, \quad \mathfrak{g} = \mathfrak{u}_1 \oplus i\mathfrak{u}_1.$$

Thus, the classification of orthogonal involutive Lie algebras of non-compact type can be further reduced to the case that $\mathfrak{g}_{\mathbb{C}}$ are still simple, and moreover, by Theorem 1, such an $(\mathfrak{g}, \sigma, Q)$ has a unique dual one $(\mathfrak{g}^*, \sigma^*, Q^*)$ of compact type, where $\mathfrak{g}^*$ is a simple compact Lie algebra classified in Lecture 6.

In this way, both the classification of symmetric spaces and that of real semi-simple Lie algebras can be naturally fitting into the classification of simple compact Lie algebras together with an involutive automorphism.

4.2. On the automorphism group of a simple compact Lie algebra and the conjugacy classes of involutive automorphisms

In this subsection, we shall use $\mathfrak{g}$ to denote a given simple compact Lie algebra and G to be the simply connected Lie group with $\mathfrak{g}$ as its Lie algebra. By Theorem 3 (Weyl) of Lecture 5, G is compact. Set $Z(G)$ to be the center of G, $\mathrm{Aut}(G) = \mathrm{Aut}(\mathfrak{g})$ to be the automorphism group of $\mathfrak{g}$ (and also that of G) and Ad to be the adjoint representation of G. Then

$$Z(G) \hookrightarrow G \xrightarrow{\mathrm{Ad}} G/Z(G) \subseteq \mathrm{Aut}(G) = \mathrm{Aut}(\mathfrak{g})$$

where $G/Z(G)$ is the identity component of $\mathrm{Aut}(\mathfrak{g})$ which is often referred to as the *group of inner automorphisms* and denoted by $\mathrm{Aut}^0(\mathfrak{g}) = \mathrm{Ad}\, G$.

The discussion of Sec. 4.1 shows that the classification of both the symmetric spaces and the real semi-simple Lie algebras can be neatly reduced to that of simple compact Lie algebras together with an involutive automorphism. In other words, the climax of such a grand classification theory actually lies in the determination of the conjugacy classes of those involutive (i.e. order 2) elements of $\text{Aut}(G)$ with respect to the adjoint action of $\text{Ad}\,G = \text{Aut}^0(\mathfrak{g})$. Technically, this is just a small step of further refinement of the structure and classification theory of compact Lie groups that we have thoroughly discussed in Lectures 3–6. Therefore, let us go back to review the results of the "compact" theory now, with such a far-reaching, wonderful specific application in mind.

(1) *The case of involutive inner automorphisms*:

Let σ be an order two element in $\text{Aut}^0(\mathfrak{g}) = \text{Ad}\,G$ and $C(\sigma)$ be the *conjugacy class* of σ in $\text{Ad}\,G$. By Theorem 1 (É. Cartan) of Lecture 3 and the surjectivity of the following maps, namely

$$\mathfrak{g} \xrightarrow{\text{Exp}} G \xrightarrow{\text{Ad}} \text{Ad}\,G = G/Z(G)$$

there exists a Cartan subalgebra $\mathfrak{h}$ of $\mathfrak{g}$ such that

$$\sigma \in \text{Ad} \cdot \text{Exp}\,\mathfrak{h} = \text{Ad}\,T = \tilde{T}, \quad \sigma^2 = \text{Id}$$

where $T = \text{Exp}\,\mathfrak{h}, \tilde{T} = \text{Ad}\,T$ are, respectively, the maximal tori of G and $\text{Ad}\,G = G/Z(G)$, namely, there exists $H \in \mathfrak{h}$ such that

$$\text{Ad} \cdot \text{Exp}\,2H = \sigma^2 = \text{Id} \Leftrightarrow \text{Exp}\,2H \in Z(G).$$

Furthermore, by the Weyl reduction (cf. Theorem 2 of Lecture 3), we may assume that H actually belongs to the fundamental domain of (W, T), namely, the closed Cartan polyhedron $\overline{P}_0$ characterized by

$$(\alpha_i, H) \geq 0, \quad \alpha_i \in \pi(\mathfrak{g}) \quad \text{and} \quad (\beta, H) \leq 1 \tag{3}$$

where β is the highest root of $\mathfrak{g}$; $\beta = \sum_{i=1}^{l} m_i \alpha_i$.

It follows from the classification of simple compact Lie algebras that the diagram of $\overline{P}_0$ together with the coefficients $\{m_i\}$ for β are as listed in Table 1, where the black dot $\{\cdot\}$ denotes the hyperplane of $\beta = 1$, while the others denote the hyperplane of $\alpha_i = 0, \alpha_i \in \pi(\mathfrak{g})$.

In retrospect, the *orbital geometry* of the adjoint action of simply connected compact Lie group that we discussed in Lecture 3, in fact, constitutes the central core of the entire structural as well as classification theory on semi-simple Lie groups and Lie algebras. The maximal tori theorem of É. Cartan and Weyl's reduction provides a clean-cut geometric structure of Ad G-action on both G (resp. $\mathfrak{g}$), namely, the Cartan's polyhedron $\overline{P}_0$ (resp. the Weyl's chamber $\overline{C}_0$) intersect every conjugacy class perpendicularly and at a unique point. First of all, such a wonderful, simple geometry on Ad G-orbital structure naturally fits perfectly well for applications of linear representation theory to analyze the structure as well as the classification theory of compact connected Lie groups (cf. Lectures 3–6), while the result can be concisely summarized in Table 1, which simply recording

Table 1

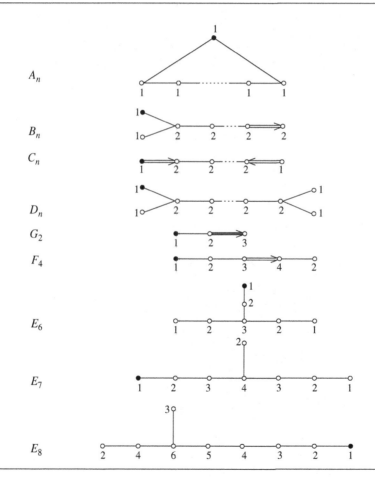

the geometry of $\overline{P}_0$ (resp. $\overline{C}_0$) of a simply connected compact Lie group (resp. its Lie algebra). Another direct application of the orbital geometry of Ad G-action is an almost straightforward proof of the Weyl's character formula, which just amounts to the computation of the volume function of principal orbits (i.e. of G/T-type).

In fact, the diagrams of Table 1 already encode many more results on the orbital geometry of Ad G-action, i.e. the assemblage of conjugacy classes. For example, let $q = \operatorname{Exp} H$ be a given point in $\overline{P}_0$ and $Z(q, G)$ be the centralizer of q in G, $Z^0(q, G)$ be its connected component of identity. Then, $Z^0(q, G) \supseteq T$ and its root system is exactly the subset of $\Delta(G)$ with $\alpha(H) \equiv 0 \pmod 1$. In particular, one has the following lemma on the center of G.

Lemma 2. *Let G be a simply connected simple compact Lie group. Then $Z(G) \subseteq \overline{P}_0$ consists of those opposite vertices of the hyperplanes in its diagram of Table 1 with label 1.*

Therefore, one has

(i) $Z(G) = 1$ (*i.e* trivial) *for G_2, F_4 and E_8.*
(ii) $Z(G) \cong \mathbb{Z}_2$ *for B_n, C_n and E_7.*
(iii) $Z(E_6) \cong \mathbb{Z}_3$, $Z(A_n) \cong \mathbb{Z}_{n+1}$.
(iv) $Z(B_{2k}) \cong \mathbb{Z}_2 \oplus \mathbb{Z}_2$, $Z(B_{2k+1}) \cong \mathbb{Z}_4$.

Corollary. *In the case $Z(G) \neq 1$, let $v_i = \operatorname{Exp} H_i$ be such a vertex and $M_i = \operatorname{Exp} \frac{1}{2} H_i$ be the middle point of the edge $\overline{ov_i}$. Then $\operatorname{Ad}(M_i)$ is an involutive automorphism of $\operatorname{Ad} G = \operatorname{Aut}^0(\mathfrak{g})$.*

Theorem 5. *Let $v_j = \operatorname{Exp} H_j$ be the vertex of $\overline{P}_0$ opposite to the hyperplane of $\alpha_j(H) = 0$ with $m_j = 2$. Then $\sigma_j = \operatorname{Ad} \operatorname{Exp} H_j$ is an involution (i.e. of order 2).*

Conversely, an involution $\sigma \in \operatorname{Ad} G$ is conjugate to either one of such σ_j, or one of that of the above corollary of Lemma 2.

Proof: Straightforward verification.   □

Let q be one of such a vertex v_j (i.e. with $m_j = 2$) or one of the kind of middle point of $\overline{ov_i}$ with $v_i \in Z(G)$, and $\sigma = \operatorname{Ad}(q)$ be its corresponding involution in $\operatorname{Aut}^0(\mathfrak{g})$. Then $\mathfrak{k} = \ker(d\sigma - \operatorname{Id})$ is exactly the Lie algebra of

$Z^0(q, G)$ and the corresponding symmetric space of compact type is just the *orbit* of q, namely,

$$G(q) = G/Z(q, G) \leftrightarrow (\mathfrak{g}, d\sigma, Q)$$

where $\mathfrak{p}$ is naturally identified with the tangent space of $G(q)$ at q equipped with the induced inner product.

Therefore, as a direct application of the above determination of involutions of $\mathrm{Aut}^0(G) = \mathrm{Ad}\, G$ and Table 1, one has the following theorem of classification of this type of irreducible compact symmetric spaces (i.e. corresponding to involutive inner automorphisms), namely

Theorem 6. *The following is the list of compact symmetric spaces corresponding to involutive inner automorphisms of compact simple Lie algebras:*

  (i) $A_n, (n \geq 1) : \mathrm{SU}(n+1)/S(U(k) \times U(n+1-k))$.
 (ii) $B_n, (n \geq 2) : \mathrm{SO}(2n+1)/\mathrm{SO}(2k) \times \mathrm{SO}(2(n-k)+1)$.
(iii) $C_n, (n \geq 3) : \mathrm{Sp}(n)/\mathrm{Sp}(k) \times \mathrm{Sp}(n-k); \mathrm{Sp}(n)/U(n)$.
 (iv) $D_n, (n \geq 4) : \mathrm{SO}(2n)/\mathrm{SO}(2k) \times \mathrm{SO}(2n-2k); \mathrm{SO}(2n)/U(n)$.
  (v) $E_6 : E_6/A_1 \times A_5; E_6/U(1) \times \mathrm{Spin}(10)$.
 (vi) $E_7 : E_7/\mathrm{SU}(8); E_7/\mathrm{SU}(2) \times \mathrm{SO}(12); E_7/U(1) \times E_6$.
(vii) $E_8 : E_8/\mathrm{Spin}(16); E_8/\mathrm{SU}(2) \times E_7$.
(viii) $F_4 : F_4/\mathrm{Spin}(9); F_4/\mathrm{SU}(2) \times \mathrm{Sp}(3)$.
 (ix) $G_2 : G_2/\mathrm{SU}(2) \times \mathrm{SU}(2)/\mathbb{Z}_2$.

Proof: $Z^0(q, G) = K$, of course, containing the maximal torus, T, of G, and moreover, the root system of $K, \Delta(K)$, consists of the subset of those roots of G with $\alpha(q) \equiv 0 \pmod{\mathbb{Z}}$. $\qquad \square$

Therefore, in the case that $q = v_j$ with $m_j = 2$, the Weyl's chamber of K is isometric to the cone of $\overline{P}_0$ at v_j, whose diagram is just that of the extended diagram of G with the dot of α_j deleted, thus having K semisimple. On the other hand, in the case that $q = M_i$, the middle point of $\overline{ov}_i$ with $m_i = 1$, K contains the circle group, $\mathrm{Exp}\, t\vec{v}_i$, as $Z^0(K)$ and the diagram of $K/Z(K)$ is just that of G with the dot of α_i deleted. Hence, the list of Theorem 6 is just the results of case by case verifications of Table 1. $\qquad \square$

(2) *The case of involutive outer automorphism*:

Let G (resp. $\mathfrak{g}$) be a given simply connected, simple compact Lie group (resp. its Lie algebra). By the fundamental theorem of Lie (cf. Lecture 2).

$$\mathrm{Aut}(G) \cong \mathrm{Aut}(\mathfrak{g}), \quad \mathrm{Aut}^0(\mathfrak{g}) = \mathrm{Ad}\, G = G/Z(G).$$

By the maximal tori theorem and the structure theorem of compact connected Lie groups (resp. simple compact Lie algebras), the induced map of an automorphism $\varphi \in \mathrm{Aut}(\mathfrak{g})$ on $\overline{C}_0$ (and hence also on $D(\mathfrak{g})$) is the identity if and only if $Q \in \mathrm{Ad}\, G = \mathrm{Aut}^0(\mathfrak{g})$, namely,

$$\mathrm{Aut}(\mathfrak{g})/\mathrm{Aut}^0(\mathfrak{g}) \to \mathrm{ISO}(D(\mathfrak{g}))$$

is injective. On the other hand, it is a corollary of Theorem 4 (Chevalley) of Lecture 5 that an isometry, p, of $D(\mathfrak{g})$ can be uniquely extended to that of $\Delta(\mathfrak{g})$ and then to an automorphism of $\mathfrak{g} \otimes \mathbb{C}$ (resp. $\mathfrak{g}$) that maps the Chevalley basis as follows, namely

$$Z_\alpha \to Z_{p(\alpha)} \quad (\text{resp. } \{X_\alpha, Y_\alpha\} \to \{X_{p(\alpha)}, Y_{p(\alpha)}\})$$

thus proving

Lemma 3. *Let $\mathfrak{g}$ be a simple compact Lie algebra. Then*

$$\iota \hookrightarrow \mathrm{Ad}\, G = \mathrm{Aut}^0(\mathfrak{g}) \hookrightarrow \mathrm{Aut}(\mathfrak{g}) \to \mathrm{ISO}(D(\mathfrak{g})) \to \iota$$

is a splitting extension of groups.

Corollary. (i) $\mathrm{Aut}(\mathfrak{g}) = \mathrm{Aut}^0(\mathfrak{g})$ *for* $\mathfrak{g} = B_n, C_n, G_2, F_4, E_7$ *and* F_8.

(ii) $\mathrm{Aut}(\mathfrak{g})/\mathrm{Aut}^0(\mathfrak{g}) \cong \mathbb{Z}_2$ *for* $\mathfrak{g} = A_n, D_n (n \neq 4)$ *and* E_6 *while* $\mathrm{Aut}(D_4)/\mathrm{Aut}^0(D_4) \cong S_3$ *(the symmetry group of* 3).

Therefore, in the study of involutive outer automorphisms, it suffices to consider the case of $\mathfrak{g} = A_n, D_n$ and E_6. We note here that $D(D_4)$ has a triple of $\mathbb{Z}_2$-elements but they are mutually conjugate, thus can be treated just the same way as the other $D_n (n \neq 4)$. Moreover, the cases of A_n and D_n are, in fact, quite simple and well-known, namely,

(i) $\mathrm{Aut}(A_n) \cong \mathrm{Ad}\, G$ extended by the addition of the involution induced by the conjugation of $\mathbb{C}$.
(ii) $\mathrm{Aut}(D_n) \cong O(2K)/\{\pm\mathrm{Id}\}$, for $n \neq 4$, while $\mathrm{Aut}(D_4)$ contains a triple of conjugating subgroups of $O(8)/\{\pm\mathrm{Id}\}$.

Therefore, for the purpose of this subsection, it remains to study the case of $\mathrm{Aut}(E_6)$ in detail.

On the structure of $\mathrm{Aut}(E_6)$

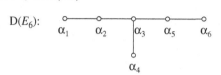

$$D(E_6):$$

Let p_0 be specific involution that permuting the pairs $\{\alpha_1, \alpha_6\}, \{\alpha_2, \alpha_5\}$ of simple roots and then accordingly the roots and their Chevalley basis, namely, starting with

$$p_0(\alpha_1) = \alpha_6, \quad p_0(\alpha_2) = \alpha_5, \quad p_0(\alpha_3) = \alpha_3, \quad p_0(\alpha_4) = \alpha_4.$$

Then, the fixed point set of p_0, say denoted by $\mathfrak{g}_0$, is a subalgebra of F_4 whose simple root system with the fixed point set of p_0 in $\mathfrak{h}$, i.e. $\mathfrak{h}_0 = \mathfrak{g}_0 \cap \mathfrak{h}$, as the Cartan subalgebra is given by

$$\alpha'_1 = \frac{1}{2}(\alpha_1 + \alpha_6), \quad \alpha'_2 = \frac{1}{2}(\alpha_2 + \alpha_5), \quad \alpha'_3 = \alpha_3, \quad \alpha'_4 = \alpha_4.$$

Now, let us analyze the orbital geometry of $\mathrm{Ad}\, E_6$-action on the connected component of $\mathrm{Aut}(E_6)$ containing p_0.

First of all, we shall denote $\mathrm{Ad}\, E_6$ simply by $\tilde{G}$ just for the sake of notational simplicity and the coset of p_0 in $\mathrm{Aut}(E_6)$ by $p_0\tilde{G}$, which is naturally equipped with a $\tilde{G}$-bi-invariant Riemannian metric. Set $\tilde{G}_{p_0}$ to be the stability (i.e. isotropy) subgroup of such an adjoint $\tilde{G}$-action, its connected component of identity is the above F_4. The restricted adjoint F_4-action on $T_{p_0}(p_0\tilde{G})$ (the tangent space of $p_0\tilde{G}$ at p_0) is isomorphic to $\mathrm{Ad}_{E_6}|F_4$ on the one hand, and on the other hand, it is the orthogonal direct sum of the isotropy representation of $\tilde{G}/F_4 = \tilde{G}(p_0)$ and the slice representation on the normal vectors of $\tilde{G}(p_0)$ at p_0. Therefore, the slice representation is isomorphic (or equivalent) to Ad_{F_4}.

Let T_1 be a chosen maximal torus of F_4. Then, the principal, isotropy subgroups of the F_4-action on the slice is (T_1), i.e. the conjugacy class of T_1 in F_4, and the same kind of proof as that of the proof maximal tori theorem in Lecture 3 will show that $p_0 F_4$ intersects every $\mathrm{Ad} - \tilde{G}$ orbit of $p_0\tilde{G}$ (i.e. non-empty).

Therefore, again by the Weyl's reduction applying to Ad_{F_4}, it is not difficult to show that the only other involutive outer automorphism of E_6 is given by $p_0 e^{\mathrm{ad}H_4/2}$ whose fixed subgroup is $C_4 \subset F_4$.

Theorem 7. *The following is the list of compact symmetric spaces corresponding to the involutive outer automorphisms of compact simple Lie algebras:*

(i) $\mathrm{SU}(k+1) : \mathrm{SU}(k+1)/\mathrm{SO}(k+1)$.
(ii) $\mathrm{SU}(2k) : \mathrm{SU}(2k)/\mathrm{Sp}(k)$.
(iii) $\mathrm{SO}(2k) : \mathrm{SO}(2k)/\mathrm{SO}(2i+1) \times \mathrm{SO}(2k-2i-1)$.
(iv) $E_6 : E_6/F_4; \; E_6/C_4$.

Proof: The case of (iv) is the result of the above analysis on outer involutions of E_6, while the cases of (i)–(iii) are the result of simple, explicit structures of $\mathrm{Aut}(A_n)$ and $\mathrm{Aut}(D_k)$ which make the classification of order 2 elements in them quite simple. □

5. Concluding Remarks

(1) The concept of Lie group structure was originated as the proper structure of group of symmetries in various mathematical structures, nowadays, often referred as *Lie transformation groups*. In retrospect, the concept of Lie groups is a natural fusion of geometric, algebraic and analytical structures that leads to a profound theory with many far-reaching applications. I think the prospects should be that there will be many more important applications of Lie group theory, especially in the realm of studying structures and problems with natural symmetries.

(2) The study of symmetric spaces was started as a bold attempt to generalize the spaces of constant curvatures, i.e. the classical geometry of Euclidean, spherical and hyperbolic spaces commonly characterized by the universality of reflectional symmetries, or equivalently, $\mathrm{ISO}(M^n, p) \simeq O(n)$ for all points of $p \in M^n$, namely, by such a drastic weakening, from $\mathrm{ISO}(M^n, p) = O(n)$ to mere $\mathrm{ISO}(M^n, p) \supseteq (\pm \mathrm{Id})$ for all points $p \in M^n$. It is truly amazing as well as a wonderful surprise that such a giant step of generalization actually leads to a neat list of classification, Cartan's monumental contribution on Lie groups and symmetric spaces. Naturally, they

constitute an outstanding family of spaces in the realm of geometry and geometric analysis. Anyhow, the geometry of symmetric spaces is naturally a rich gold mine for geometers, with abundance of natural but challenging problems, especially in the realm of profound interplays between symmetric properties and various basic geometric objects.

(3) Let $M = G/K$ be a symmetric space, namely, $G = \mathrm{ISO}(M)$ and $K = \mathrm{ISO}(M, p)$. Then, the *orbital geometry* of the K-action on M can be regarded as the localized symmetry property of M. Note that a compact connected Lie group G_1 (with bi-invariant metric) is, itself, a symmetric space, namely, with

$$G = \mathrm{ISO}(M) = G_1 \times G_1 / \Delta Z(G_1) \quad \text{and} \quad K = \mathrm{ISO}(M, \mathrm{Id}) = \Delta(G_1 / Z(G_1)).$$

In such a special case, the K-action on M (i.e. G_1) is exactly the adjoint action of G_1. Therefore, the orbital geometry of K-action on M is a natural generalization of the orbital geometry of Ad_{G_1}-action on G_1, and hence, what we are looking for is a generalization of maximal tori theorem and Weyl's reduction in the context of symmetric spaces.

(4) *Morse theory on symmetric spaces [B-S]*: Geodesics are clearly the simplest and the most basic geometric objects on Riemannian manifolds in general. The Morse theory of geodesics studies the global behaviors of geodesics, while the fundamental equation is the Jacobi's equation along geodesics. In the special case of symmetric spaces, the Jacobi's equation becomes ODE of constant coefficients because of the existence of *transvections* along geodesics in symmetric spaces, thus making the analysis of nonuniqueness of shortest pathways in the case of compact symmetric spaces much simpler than otherwise. Such analysis in [B-S] shows that Jacobi vector fields which *contribute to focal points* are already provided by Killing vector fields (this is exactly the *variational completeness* of [B-S]), thus making the orbital geometry of K-action on a compact symmetric space M, a powerful tool in the analysis of Morse theory of geodesics. Furthermore, such study naturally led Bott to the discovery of Bott's periodicity [B].

(5) Observe that geodesics are optimal geometric objects of 1-dimensional variational problem, while *isoperimetric regions* are the optimal geometric objects of the top dimensional variational problem. Let us conclude our remarks by the following problem on isoperimetric regions in *non-compact* symmetric spaces, namely,

Problem Let $M = G/K$ be a non-compact symmetric space and $\Omega \subset M$ is an isoperimetric region. Is Ω necessary K-invariant?

Suppose that the answer of the above problems turns out to be affirmative in general. That will be a spectacular generalization of the classical isoperimetric problem that plays a central role on geometric variation theory dated way back to antiquity Greek geometry in the simplest case of $\mathbb{E}^2$.

On the other hand, if it holds for certain non-compact symmetric spaces but no longer holding in general, i.e. it needs certain modification. Then such theorems would be even more interesting and inspiring toward the deeper understanding of the interplay between symmetry and optimization.

Printed in the United States
By Bookmasters